Principles of Chemical Kinetics

J. E. House

Department of Chemistry
Illinois State University

WCB **Wm. C. Brown Publishers**

Dubuque, IA Bogota Boston Buenos Aires Caracas Chicago Guilford, CT
London Madrid Mexico City Seoul Singapore Sydney Taipei Tokyo Toronto

Book Team

Developmental Editor   *Brittany J. Rossman*
Publishing Services Coordinator   *Julie Avery Kennedy*

 **Wm. C. Brown Publishers**

President and Chief Executive Officer   *Beverly Kolz*
Vice President, Publisher   *Kevin Kane*
Vice President, Director of Sales and Marketing   *Virginia S. Moffat*
Vice President, Director of Production   *Colleen A. Yonda*
National Sales Manager   *Douglas J. DiNardo*
Marketing Manager   *Kristin M. Foley*
Advertising Manager   *Janelle Keeffer*
Production Editorial Manager   *Renée Menne*
Publishing Services Manager   *Karen J. Slaght*
Royalty/Permissions Manager   *Connie Allendorf*

A Times Mirror Company

Copyediting and production by Shepherd, Inc.

Library of Congress Catalog Card Number: 95-83493

ISBN 0-697-32881-3

Printed in the United States of America by Times Mirror Higher Education Group, Inc.
2460 Kerper Boulevard, Dubuque, IA 52001

10   9   8   7   6   4   3   2   1

To Kathy

CONTENTS

PREFACE   ix

1   SOME FUNDAMENTAL IDEAS OF KINETICS   1

1.1	Rates of Reactions	2
1.2	Dependence on Concentration	4
	1.2.1 First-Order	4
	1.2.2 Second-Order	7
	1.2.3 Zero-Order	9
1.3	Cautions on Treating Kinetic Data	11
1.4	Effects of Temperature	14
1.5	Mechanisms of Reactions	18
	1.5.1 Direct Combination	19
	1.5.2 Chain Mechanisms	19
	1.5.3 Substitution Reactions	21
1.6	Catalysis	24

2   KINETICS OF MORE COMPLEX SYSTEMS   31

2.1	Second-Order Case, First-Order in Two Components	31
2.2	Other Reaction Orders	35
2.3	Parallel First-Order Reactions	38
2.4	Series First-Order Reactions	40
2.5	Reversible Reactions	45
2.6	Autocatalysis	51
2.7	Effect of Temperature	56

3   TECHNIQUES AND METHODS   65

3.1	Calculating Rate Constants	65
3.2	Method of Half-Lives	66
3.3	Initial Rates	68
3.4	Flooding	70
3.5	Logarithmic Method	71
3.6	Effects of Pressure	73
3.7	Flow Techniques	77
3.8	Tracer Methods	78
3.9	Kinetic Isotope Effects	81

4 REACTIONS IN THE GAS PHASE 89

4.1	Collision Theory	89
4.2	The Potential Energy Surface	93
4.3	Transition State Theory	98
4.4	Unimolecular Decomposition of Gases	103
4.5	Free-Radical and Chain Mechanisms	110
4.6	Adsorption	115
	4.6.1 Langmuir Adsorption Isotherm	117
	4.6.2 B-E-T Isotherm	121
	4.6.3 Poisons and Inhibitors	122
4.7	Catalysis	123

5 REACTIONS IN SOLUTIONS 131

5.1	The Nature of Liquids	131
	5.1.1 Intermolecular Forces	132
	5.1.2 The Solubility Parameter	136
	5.1.3 Solvation of Ions and Molecules	139
	5.1.4 The Hard-Soft Interaction Principle	140
5.2	Solvent Polarity Effects on Rates	142
5.3	Ideal Solutions	144
5.4	Cohesion Energies of Ideal Solutions	147
5.5	Effects of Solvent Cohesion Energy on Rates	150
5.6	Solvation and Its Effects on Rates	151
5.7	Effects of Ionic Strength	155
5.8	Linear Free-Energy Relationships	158
5.9	The Compensation Effect	162
5.10	Some Correlations of Rate with Solubility Parameter	163

6 ENZYME CATALYSIS 175

6.1	Enzyme Action	176
6.2	Kinetics of Reactions Catalyzed by Enzymes	178
	6.2.1 Michaelis-Menten Analysis	178
	6.2.2 Lineweaver-Burk and Eadie Analyses	183
6.3	Inhibition of Enzyme Action	185
	6.3.1 Competitive Inhibition	186
	6.3.2 Noncompetitive Inhibition	187
	6.3.3 Uncompetitive Inhibition	189
6.4	Enzyme Activation by Metal Ions	191
6.5	Regulatory Enzymes	192

7 KINETICS OF REACTIONS IN THE SOLID STATE 195

7.1 General Considerations 195
7.2 Factors Affecting Reaction Rates in Solids 198
7.3 Rate Laws 199
    7.3.1 The Parabolic Rate Law 199
    7.3.2 The First-Order Rate Law 200
    7.3.3 The Contracting Sphere Rate Law 201
    7.3.4 The Contracting Area Rate Law 202
7.4 The Prout-Tompkins Equation 205
7.5 Rate Laws Based on Nucleation 207
7.6 Kinetic Studies 210
    7.6.1 The Deaquation-Anation of $[Co(NH_3)_5H_2O]Cl_3$ 212
    7.6.2 The Deaquation-Anation of $[Cr(NH_3)_5H_2O]Br_3$ 215
    7.6.3 The Dehydration of *Trans*-$[Co(NH_3)_4Cl_2]IO_3 \cdot 2H_2O$ 216

8 NONISOTHERMAL METHODS IN KINETICS 221

8.1 TGA and DSC Methods 221
8.2 Kinetic Analysis by the Coats and Redfern Method 224
8.3 The Reich and Stivala Method 227
8.4 A Method Based on Three (α,T) Data Pairs 228
8.5 A Method Based on Four (α,T) Data Pairs 230
8.6 A Differential Method 231
8.7 A Comprehensive Nonisothermal Kinetic Method 232
8.8 The General Rate Law and Comprehensive Method 233

INDEX 239

PREFACE

Principles of Chemical Kinetics was developed for a two-credit course in chemical kinetics for first-year graduate students and, occasionally, some advanced undergraduates. Examination of existing texts revealed what I perceive to be serious omissions in the coverage of topics.

The course I had planned was a *balanced* course that would include most of the important areas of kinetics. I believe that a course in kinetics should begin with a review of basic concepts, presentation of the mathematical analysis of rate laws, and a description of the various techniques and methods used in the study of reactions. All kinetics texts do begin that way. Following that, the important areas include gas phase reactions, reactions in solutions, enzyme kinetics, and reactions in solids. An examination of the available books on kinetics showed that most of them are specialized and deal in great detail with only a portion of the field. The books that deal with gas phase reactions may cover no enzyme kinetics; books dealing with enzyme kinetics may not treat reactions in the gas phase; none cover reactions in solids, etc. There being no single text dealing with the several areas of reaction kinetics, the present book was conceived.

This book is intended to serve as a *text* for students wishing to learn (formally or informally) the essentials of chemical kinetics as applied to the broad spectrum of chemistry. My attempt has been to present chemical kinetics in a clear and unified way. The topics included have been chosen in order to make the book useful to students in all areas of chemistry, biochemistry, and, perhaps, materials science while not becoming overly laborious. At the outset, it was recognized that the book must not become encyclopedic, but rather it must conform to rather severe size limitations. The chapter on reactions in solids presents material not available in any of the currently available kinetics texts. A brief chapter on nonisothermal kinetics methods is included because of the increasingly important role of such methods, especially in industrial chemistry.

Chapter 1 is intended to provide a review of basic concepts or an elementary survey for persons studying kinetics for the first time. After covering Chapters 1 to 3, the remaining chapters can be covered in any order because they essentially stand alone as descriptions of the topics. Also, omission of one of the chapters does not render subsequent topics unintelligible. This is not a book that is designed to take the reader to the frontiers of the theory of chemical kinetics. It is meant to be a practical guide to the underlying principles that form the basis of kinetic studies of all types. Undoubtedly, there will be those who believe that more coverage is desirable on certain topics. More specialized books on the various aspects of kinetics exist. This book is designed to provide a reasonable presentation of several areas of chemical kinetics.

It is hoped that this book will be of use to graduate students who are beginning the study of chemical kinetics and to undergraduate students who wish to supplement their knowledge of topics in kinetics that may have been introduced in undergraduate courses. The book could be the primary text in some courses or it could be an ancillary or "lead-in" text in kinetics courses that focus on a specific topic (e.g., gas phase kinetics). It is also hoped that the book might be used by practicing chemists who need to review or learn the essentials of kinetics through self-study. In order to be of use to the widest audience, the book begins with a review of elementary concepts before progressing to the main body of information. The aim has been to provide sufficient material to serve as the background for those undertaking further study of specific areas of kinetics and to provide a survey of chemical kinetics for all chemists. It is hoped that this book meets those objectives. Certainly its broad content in so small a volume renders it unique among books on chemical kinetics.

The author wishes to thank those persons who have made the writing, development, and production of this book a pleasant and rewarding experience. Thanks to the students in Chemistry 464, Chemical Kinetics, who used an earlier version of this material and made many useful suggestions. To Drs. Earl Pearson, Clarke Earley, and Mark Gordon go my thanks for critically reviewing the manuscript and indicating many ways to improve it. Thanks to John Berns, Julie Kennedy, Brittany Rossman, Molly Kelchan, and Jeff Hahn of Wm. C. Brown Publishers for their help at all levels. Finally, the author wishes to thank his wife, Dr. Kathleen A. House, for her help in using computer graphics software and for her encouragement in this project from its inception. Her suggestions and meticulous reading of the proofs have greatly improved the final product.

chapter 1

SOME FUNDAMENTAL IDEAS OF KINETICS

It is ordinarily observed that the majority of reactions that lead to a lower over-all energy state take place readily. However, there are many reactions that lead to a decrease in energy yet the rates of the reactions are low. For example, the heat of formation of water is –285 kJ/mol, but the reaction

$$H_2(g) + 1/2\,O_2(g) \longrightarrow H_2O(l) \qquad (1.1)$$

takes place very slowly, if at all, unless the reaction is started by a spark. The reason for this is that although a great deal of energy is released, there is no low-energy pathway for the reaction to follow. In order for water to form, molecules of H_2 and O_2 must react, and their bond energies are about 435 and 490 kJ/mol, respectively.

Thermodynamics is concerned with the overall energy change for a process. If necessary, this change can result after an infinite time. Accordingly, thermodynamics does not deal with the subject of rates of reactions, at least not directly. The example above shows that the thermodynamics of the reaction favors the production of water; however, kinetically the process is unfavorable. We see here the first of several important principles of chemical kinetics. There is no necessary correlation between thermodynamics and kinetics. Some reactions that are energetically favorable take place very slowly because there is no low-energy pathway for the reaction.

One of the features of the study of reaction rates is that a rate can not be calculated. Theory is not developed to the point where it is possible to calculate how fast most reactions will take place. For some very simple gas phase reactions, it is possible to calculate approximately how fast the reaction should take place, but details of the process must still be determined experimentally.

1.1 RATES OF REACTIONS

The rate of a chemical reaction is expressed as a change in concentration of some species with time. Therefore, the units of the *rate* must be those of concentration/time (moles/liter sec, moles/liter min, etc.). A reaction such as

$$A \longrightarrow B$$

has a rate which can be expressed either in terms of the disappearance of A or the appearance of B. Because the concentration of A is decreasing as A is consumed, the rate is expressed as $-d[A]/dt$. Because the concentration of B is increasing with time, the rate is expressed as +(change in [B]/change in time) or $+d[B]/dt$. The actual mathematical equation relating concentrations and rate is called the *rate equation* or *rate law*. The relationship between the change in concentration of [A] and [B] is represented in Figure 1.1.

If we consider a reaction such as

$$a A + b B \longrightarrow c C + d D \tag{1.2}$$

the rate law will usually be expressed in terms of a constant times some function of concentrations, and it can be written in the form

$$\text{Rate} = k[A]^x[B]^y \tag{1.3}$$

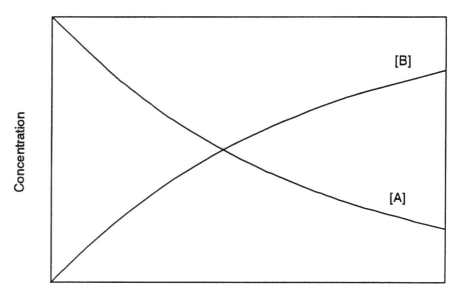

Figure 1.1 Change in concentration of A and B as A $\longrightarrow$ B.

where x and y are the exponents on the concentrations of A and B respectively. In this rate law, k is called the *rate constant* and the exponents x and y are called the *order* of the reaction with respect to A and B, respectively. As will be described later, the exponents x and y may or may not be the same as the balancing coefficients a and b in Eq. (1.2). The overall order of the reaction is the sum of the exponents x and y. Thus, we speak of a second-order reaction, a third-order reaction, etc. These exponents must be established by studying the reaction using different initial concentrations of A and B. When this is done, it is possible to determine if doubling the concentration of A doubles the rate of the reaction. If it does, then the reaction must be first-order in A, and the value of x is 1. However, if doubling the initial concentration of A quadruples the rate, it is clear that [A] must have an exponent of 2, and the reaction is second-order in A. One very important point to remember is that there is no *necessary* correlation between the balancing coefficients in the equation and the exponents in the rate law. They may be the same but one can not assume that they will be without studying the rate of the reaction.

A study of the rate of a reaction gives information about the slowest step of the reaction if it takes place in a series of steps. We can see an analogy to this in the following example.

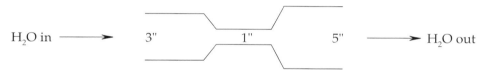

H_2O in $\longrightarrow$ 3"   1"   5" $\longrightarrow$ H_2O out

If we study the flow of water through this system, we will get information about the flow of water through a 1" pipe since the 3" and 5" pipes do not limit the flow of water. Therefore, in the language of chemical kinetics, the 1" pipe is the *rate-determining step*.

Suppose we have a reaction which can be written as

$$2A + B \longrightarrow \text{Products} \tag{1.4}$$

and that the reaction takes place in the steps shown as

$$A + B \longrightarrow C \qquad \text{(slow)} \tag{1.5}$$

$$C + A \longrightarrow \text{Products} \qquad \text{(fast)} \tag{1.6}$$

The amount of the intermediate C produced limits the rate of the overall reaction. Note that the sum of Eqs. (1.5) and (1.6) gives the overall reaction shown in Eq. (1.4). Note also that the formation of C depends on the reaction of one molecule of A and one of B. That process will then have a rate that is dependent on $[A]^1$ and $[B]^1$. Therefore, even though the balanced overall equation involves two molecules of A, the *slow step* involves only one molecule of A. Formation of products then follows a rate law that is of the form Rate = $k[A][B]$ and the reaction is second-order (first-order in [A] and first-order in [B]). It should be apparent that we can write the rate law directly from the balanced equation only if the reaction takes place in a single step. If the reaction takes place in a series of steps, a rate study will give information about steps up to and including the slowest step, and the rate law will be determined by that step.

1.2 DEPENDENCE ON CONCENTRATION

In this section, we will examine the details of rate laws that depend on concentration of reactants in some simple way. While many complicated cases are well known, there are also a great many reactions for which the dependence on concentration is first-order, second-order, or zero-order.

1.2.1 First-Order

Suppose a reaction can be written as

$$A \longrightarrow B \tag{1.7}$$

and the rate law can be written as

$$\text{Rate} = k[A]^1 = -\frac{d[A]}{dt}$$

We can now write

$$-\frac{d[A]}{[A]} = k \, dt$$

This equation can be integrated but it should be integrated between the limits of time = 0 and time equal to some later time, t, while the concentration varies from $[A]_o$ at time zero to $[A]$ at the later time. This is written as

$$-\int_{[A]_o}^{[A]} \frac{d[A]}{[A]} = k \int_0^t dt \tag{1.8}$$

which gives, when the integration is performed,

$$\ln \frac{[A]_o}{[A]} = kt \quad \text{or} \quad \log \frac{[A]_o}{[A]} = \frac{k}{2.303} t \tag{1.9}$$

If the equation involving natural logarithms is considered, it can be written as

$$\ln[A]_o - \ln[A] = kt$$

or

$$\ln[A] = \ln[A]_o - kt \qquad (1.10)$$
$$y = b + mx$$

It must be remembered that $[A]_o$, the initial concentration of A, is a constant. Therefore, Eq. (1.10) can be put in the form of a straight line with $y = \ln[A]$, $m = -k$, and $b = \ln[A]_o$. A graph of $\ln[A]$ versus time will be linear with a slope of $-k$. In order to test this rate law, it is necessary to have data for the reaction which consists of the concentration of A determined as a function of time. This suggests that in order to carry out a kinetic study, one must have an analytical method to determine the concentration of some species. Simple, reliable, and rapid analytical methods are usually sought. Additionally, one must measure time, which is not usually a problem unless the reaction is a very rapid one. What is usually done is to set up the reaction in a constant temperature bath so that fluctuations in temperature will not cause changes in rate. Then, the reaction is started, and the concentration of the reactant (A in this case) is determined at selected times so that a graph of $\ln[A]$ versus time can be made. If a straight line results, the reaction is following a first-order rate law. Graphical representation of this rate law is shown in Figure 1.2. In this case, the slope of the line is $-k$, so the experimental data can be used to determine k graphically.

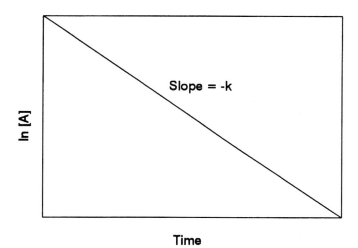

Figure 1.2 First-order plot for A ⟶ B.

The units on k in the first-order rate law are in terms of time^{-1}. The left-hand side of Eq. (1.9) has [concentration]/[concentration] so that the units cancel. However, the right-hand side of the equation will be correct only if k has the units of time^{-1}, because only then kt will have no units.

The equation

$$\ln[A] = \ln[A]_o - kt$$

can also be put in the form

$$[A] = [A]_o e^{-kt} \tag{1.11}$$

From this equation, it can be seen that the concentration of A decreases with time in an exponential way. Such a relationship is sometimes referred to as an *exponential decay*.

Radioactive decay processes follow a first-order rate law. Doubling the amount of radioactive material doubles the rate of decay. When the amount of material remaining is one-half of the original amount, the time expired is called the *half-life*. We can calculate it easily using Eq. (1.9). At the point where time is one half-life, $t = t_{1/2}$, the concentration of A is one-half the initial concentration, $[A]_o/2$. Therefore, we can write

$$\ln \frac{[A]_o}{[A]} = \ln \frac{[A]_o}{[A]_o/2} = kt_{1/2} = \ln 2 = 0.693$$

The half-life is then given as

$$t_{1/2} = 0.693 / k \tag{1.12}$$

and it will have units which depend on the units on k. For example, if k is hr^{-1}, then the half-life is in hours, etc. Note that for a process following a first-order rate law, the half-life is independent of the initial concentration of reactant. For radioactive decay, the half-life is independent of the amount of starting material. For example, if a sample initially contains 1000 atoms of radioactive material, the half-life is exactly the same as when there are 5000 atoms initially present.

It is easy to see that after one half-life, the amount of material remaining is one-half of the original; after two half-lives, the amount remaining is one-fourth of the original; after three half-lives, the amount remaining is one-eighth of the original, etc. This is illustrated graphically in Figure 1.3.

While the term *half-life* might more commonly be applied to radioactivity, it is just as appropriate to speak of the half-life of a chemical reaction as the time for the concentration of some reactant to fall to one-half of its initial value. We will have occasion to return to this point.

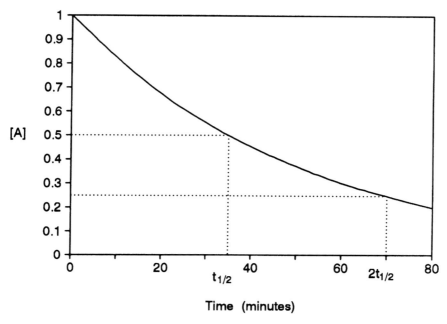

Figure 1.3 Half-life determination for a first-order process.

1.2.2 Second-Order

A reaction that is second-order in one component obeys the rate law

$$\text{Rate} = k\,[A]^2 = -\frac{d[A]}{dt} \tag{1.13}$$

Such a rate law *might* result from a reaction such as

$$2A \longrightarrow \text{Products} \tag{1.14}$$

but, as we have seen, the rate law can not be written from the balanced equation. If we rearrange Eq. (1.13), we have

$$-\frac{d[A]}{[A]^2} = k\,dt$$

If the equation is integrated between limits on concentration of $[A]_o$ at $t = 0$ and $[A]$ at time t, we have

$$-\int_{[A]_o}^{[A]} \frac{d[A]}{[A]^2} = k\int_0^t dt$$

which gives the integrated rate law

$$\frac{1}{[A]} - \frac{1}{[A]_o} = kt \tag{1.15}$$

Since the initial concentration of A is a constant, the equation can be put in the form

$$\frac{1}{[A]} = kt + \frac{1}{[A]_o} \tag{1.16}$$

$$y = mx + b$$

As shown in Figure 1.4, a plot of $1/[A]$ versus time should be linear with a slope of k and an intercept of $1/[A]_o$ if the reaction follows the second-order rate law.

The half-life for a reaction following the second-order rate law can be easily calculated. At time equal to one half-life, the concentration of A has decreased to one-half its original value. That is, $[A] = [A]_o/2$ so that

$$\frac{1}{[A]_o/2} - \frac{1}{[A]_o} = kt_{1/2} \tag{1.17}$$

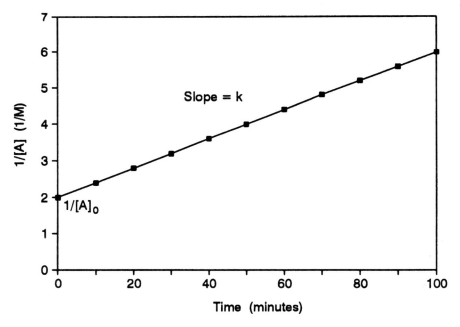

Figure 1.4 Second-order rate plot where $[A]_o$ is 0.5 M and $k = 0.040$ liter/mol min.

or

$$\frac{2}{[A]_o} - \frac{1}{[A]_o} = kt_{1/2} = \frac{1}{[A]_o}$$

Therefore,

$$t_{1/2} = 1/k[A]_o \qquad (1.18)$$

Here we see a major difference between a reaction that follows a second-order and one that follows a first-order rate law. In the first-order case, the half-life is independent of the initial concentration of the reactant, but in the second-order case, the half-life is inversely proportional to the initial concentration of the reactant.

1.2.3 Zero-Order

For certain reactions, the rate is independent of the concentration of reactant over a wide range of concentration. For example, the decomposition of hypochlorite on a cobalt oxide catalyst behaves this way. The reaction is

$$2OCl^- \longrightarrow 2Cl^- + O_2 \qquad (1.19)$$

The cobalt oxide catalyst forms when a solution containing Co^{2+} is added to the solution containing the OCl^-. It is likely that some of the cobalt is oxidized to Co^{3+} so we will write the catalyst as Co_2O_3 even though it is probable that some CoO is also present.

The reaction takes place on the surface of the solid particles. This happens because OCl^- is adsorbed to the solid and the surface becomes essentially covered. Thus, the concentration of OCl^- in the solution does not matter as long as there is enough to cover the surface of the catalyst. What does matter in this case is the surface area of the catalyst. As a result, the decomposition of OCl^- on a given fixed amount of catalyst occurs at a constant rate over a wide range of OCl^- concentrations.

For a reaction in which a reactant disappears in a zero-order process, we can write

$$-\frac{d[A]}{dt} = k[A]^0 = k \qquad (1.20)$$

since $[A]^0 = 1$. Therefore, we can write

$$-d[A] = k \, dt$$

or

$$-\int_{[A]_o}^{[A]} d[A] = k \int_0^t dt$$

Integration of this equation between the limits of $[A]_o$ at zero time and $[A]$ at some later time, t, gives

$$[A] = [A]_o - kt \qquad \qquad (1.21)$$

This equation indicates that at any time after the reaction starts, the concentration of A is the initial value minus a constant times t. This equation can be put in the form of a straight line.

$$[A] = -k \cdot t + [A]_o$$
$$y = m \cdot x + b$$

Figure 1.5 shows such a graph for a process that follows a zero-order rate law and the slope of the line is $-k$.

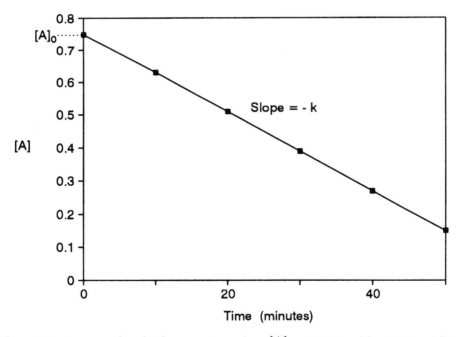

Figure 1.5   A zero-order plot for a reaction where $[A]_o = 0.75$ **M** and $k = 0.012$ mol/liter min.

As in the previous cases, we can determine the half-life because after one half-life, $[A] = [A]_o/2$. Therefore,

$$[A]_o / 2 = [A]_o - k t_{1/2}$$

so that

$$t_{1/2} = \frac{[A]_o}{2k} \qquad \textbf{(1.22)}$$

In this case, we see that the half-life is directly proportional to $[A]_o$, the initial concentration of A.

While this type of rate law is not especially common, it is followed by some reactions, usually ones in which some other factor governs the rate. This is the case for the decomposition of OCl^- described above. An important point to remember for this type of reaction is that *eventually* the concentration of OCl^- becomes low enough that there is not a sufficient amount to replace quickly that which reacts on the surface of the catalyst. Therefore, the concentration of OCl^- does limit the reaction in that case, and the reaction is no longer independent of $[OCl^-]$. The rate of reaction is independent of $[OCl^-]$ over a wide *range* of concentrations, but it is not *totally* independent of $[OCl^-]$. The reaction is not strictly zero-order, but it *appears* to be so. Such a reaction is said to be *pseudo* zero-order. This situation is similar to reactions in aqueous solutions in which we treat the concentration of water as being a constant when a slight amount of it reacts. We treat the concentration as being constant because the amount reacting compared to the amount present is very small. We will describe other pseudo-order processes in a later section.

1.3   CAUTIONS ON TREATING KINETIC DATA

It is important to realize that when graphs are made to fit data to the rate laws, the points are not without some experimental error. Typically, the larger part of the error is in the analytical measurement, and the smaller part is in the measurement of time. In order to show some aspects which can affect the interpretation of data, consider the case illustrated in Figure 1.6. In this case, we must decide how to draw the line giving the best fit to the data. One method is to simply visually decide where the line should be drawn so that it passes closest to the largest number of points. This method, although rapid, is not the most accurate method. A better way is to fit the points using linear regression or least squares. In this method, a calculator or computer is used to calculate the sums of the squares of errors and then the "line" is established which makes these sums a minimum. This mathematical method removes the necessity of drawing

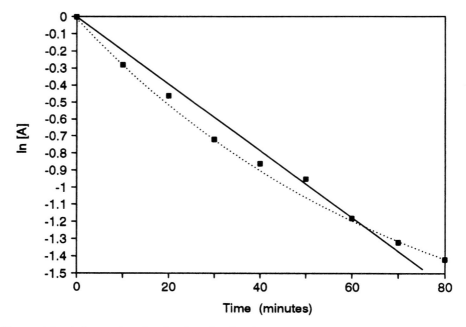

Figure 1.6 A first-order plot for data that has large errors.

the line at all since the slope, intercept, and correlation coefficient (a statistical measure of "goodness" of fit) are determined.

While the above procedures are straightforward, there may still be some problems. For example, suppose for some reaction A ———→ B, we determine the following data.

Time (min)	[A]	ln[A]
0	1.00	0.00
15	0.86	−0.151
30	0.80	−0.223
45	0.68	−0.386
60	0.57	−0.562

If we plot these data to test the zero- and first-order rate laws, we obtain the graphs shown in Figure 1.7. It is easy to see that the two graphs give about equally good fits to the data. Therefore, on the basis of the data above, it would not be possible to say whether the reaction is zero- or first-order.

What has happened in this case is that the errors in the data points have made it impossible to decide between a curve and a straight line. What we are

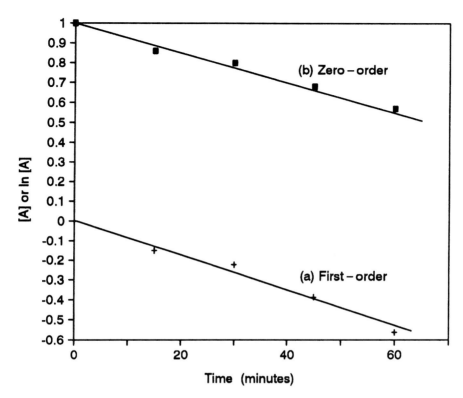

Figure 1.7   Rate plots for data shown in text.

examining is only a small portion of the total curve showing concentration versus time. In a narrow range of one of the variables, almost any mathematical function will represent the curve fairly well. The way around this is to study the reaction over several half-lives so that the dependence on concentration can be determined. Only the correct rate law will represent the data when a large extent of the curve is considered. The fundamental problem is one of distinguishing between the two cases shown in Figure 1.7.

In Figure 1.6, one function shown represents an incorrect rate law, while the other represents the correct rate law but with rather large errors in the data. Clearly, to insure that a kinetic study is properly carried out, the experiment should be repeated several times and it should be studied over a sufficient range of concentration so that any errors will not make it impossible to determine which rate law is the best fitting one. After the correct rate law is identified, several runs can be carried out so that an average value of the rate constant can be determined.

1.4 EFFECTS OF TEMPERATURE

In order for molecules to be transformed from reactants to products, it is necessary that they pass through some energy state that is higher than either the reactants or products. For example, it might be necessary to bend or stretch some bonds in the reactant molecule before it is transformed into a product molecule. Although other cases will be discussed in later sections, the essential idea here is that some state of higher energy must be populated as shown in Figure 1.8. Such a situation immediately suggests the *Boltzmann Distribution Law* as providing the explanation. In this case, [ ]* denotes the high-energy state, which is called the *transition state* or the *activated complex*. The case illustrated represents an exothermic reaction since the overall energy change is negative with the products having a lower energy than the reactants.

When the rate laws are inspected, we see that only k can be a function of T because the concentration remains constant or nearly so as the temperature changes. Therefore, it is the rate constant which incorporates information about the effect of temperature on the rate of a reaction.

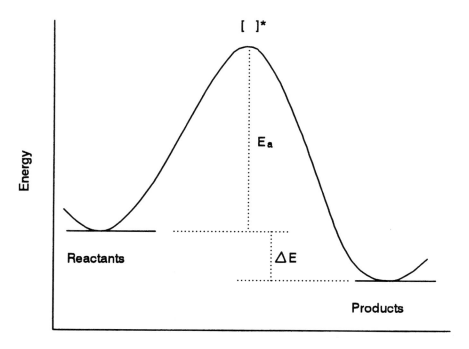

Figure 1.8 Energy profile for a chemical reaction.

There are several types of behavior exhibited as one studies the rates of reactions as a function of temperature. Three of the most common variations in rate are shown in Figure 1.9. The first shows the variation followed by most reactions, that of an increasing rate as temperature increases. The second shows the behavior of some material that becomes explosive at a certain temperature. At temperatures below the explosive limit, the rate is essentially unaffected by temperature. Then, as the material becomes explosive, the rate increases enormously at some temperature where the explosion begins. In the third case, we see the variation in rate of reaction for many biological reactions. For example, reactions involving enzymes (biological catalysts) frequently increase in rate up to some temperature and then decrease in rate at higher temperatures. Enzymes are protein materials which become denatured at high temperatures, so their reactions usually have some optimum temperature where the rate is a maximum (see Chapter 6).

Arrhenius suggested in the late 1800s that the rate of most reactions (Figure 1.9 a) varies with temperature in such a way that

$$k = Ae^{-E_a/RT} \tag{1.23}$$

where k is the rate constant, A is the frequency factor or pre-exponential factor, and T is the temperature (K). If we take the natural logarithm of both sides of Eq. (1.23) we obtain

$$\ln k = -\frac{E_a}{RT} + \ln A \tag{1.24}$$

This equation can be put in the form of a straight line,

$$\ln k = -\frac{E_a}{R} \cdot \frac{1}{T} + \ln A \tag{1.25}$$
$$y = m \cdot x + b$$

Therefore, a plot of $\ln k$ versus $1/T$ can be made after the rate constant has been determined for the reaction carried out at several temperatures. The slope of the line is $-E_a/R$, and the intercept is $\ln A$. Such a graph, like the one shown in Figure 1.10, is often called an *Arrhenius plot*.

In determining the activation energy from an Arrhenius plot, it is important to observe several precautions. For example, if the reaction is being studied at 300 K, $1/T$ will be 0.00333 deg^{-1}. If the reaction is then studied at 305 K, $1/T$ will be 0.00328 deg^{-1}. Such a small difference in $1/T$ makes it very difficult to determine the slope of the line accurately, especially if the temperature has not been controlled *very* accurately. Consequently, it is desirable to study a reaction over as large a range of temperature as possible and to use several temperatures in that range in order to minimize errors. For most reactions, 20 to 25° represents

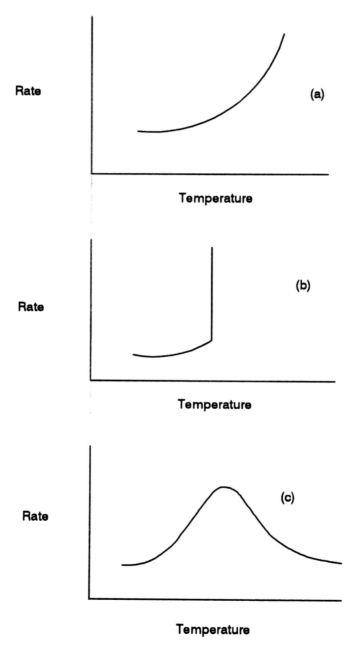

Figure 1.9   Some of the ways in which reactions rates vary with temperature.

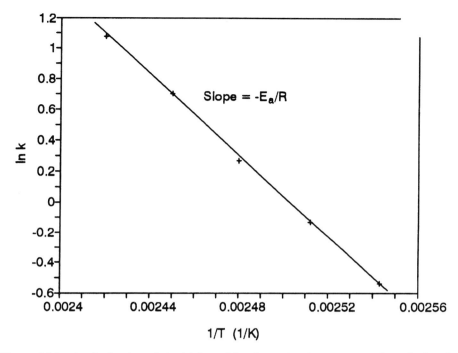

Figure 1.10 An Arrhenius plot which enables the activation energy to be calculated.

the range in which the reaction can be studied because at low temperatures, the reaction is very slow and at higher temperatures, the reaction is very fast. Therefore, it is generally desired to study the reaction over a range of about 20 to 25°.

If a reaction is studied at only two temperatures, it is still possible to evaluate the activation energy, but such a case is not nearly as desirable as the procedure described above. If k_1 is the rate constant at T_1, and k_2 is the rate constant at T_2, we can write

$$\ln k_1 = \ln A - E_a / RT_1$$

$$\ln k_2 = \ln A - E_a / RT_2$$

Subtracting the equation for $\ln k_2$ from that giving $\ln k_1$ gives

$$\ln k_1 - \ln k_2 = (\ln A - E_a / RT_1) - (\ln A - E_a / RT_2)$$

We can simplify the equation to obtain

$$\ln\frac{k_2}{k_1} = \frac{E_a\,(T_2 - T_1)}{RT_1 T_2} \tag{1.26}$$

To carry out a kinetic study of a reaction, the following things must be done as a minimum before the reaction can be described kinetically.

1. Run the reaction at a fixed temperature in a constant temperature bath and determine the concentration of a reactant or product after various time intervals.
2. Fit the data to a rate law keeping in mind that the reaction must be studied over several half–lives and the experiment should be carried out several times.
3. Determine the rate constant at the temperature for which the reaction was studied. An average value of k from several runs is preferred.
4. After the rate law is known, study the reaction over as wide a range of temperature as possible, repeating steps 1–3 above. Make replicate runs at each temperature.
5. After the average rate constant is obtained at each temperature, prepare an Arrhenius plot and determine the activation energy from the slope.

This is certainly a simplified kinetic study and other factors may have to be investigated in other cases. For example, the effect of changing the solvent for the reaction is frequently studied. Also, the presence of materials that do not participate directly in the reaction may affect the rate of the reaction. These and other factors may be studied in particular cases.

1.5 MECHANISMS OF REACTIONS

When using the term *mechanism* for a reaction, we mean the details of the number of molecules and their arrangement at the time the reaction occurs. A rate law gives the *molecularity* of the reaction so we can find out how many molecules are involved in forming the transition state from the rate law. Frequently, other experiments are required to determine other information about the reaction. We will see examples of this when specific reactions are described.

Some reactions appear to occur as a direct result of molecular collision, especially for reactions in the gas phase. However, it is not simple to calculate the number of collisions, the fraction of collisions having great enough energy to form products, and the fraction of collisions which have the molecules in exactly the right orientation to form products. As a result, reaction rates must be measured even for gaseous reactions. For reactions taking place in solutions, the factors described previously are important but there are also effects caused by the solvent. For example, if a reactant is polar or ionic, it will be strongly solvated in a polar solvent such as water. Also, in aqueous solutions, there will be the effects of acidity or basicity to consider. Even with all these problems, there have been many reactions studied in sufficient detail that the mechanisms are fairly well understood. We will now describe a few cases which will serve to illustrate the general approaches used to study mechanisms.

1.5.1 Direct Combination

The reaction between $H_2(g)$ and $I_2(g)$ has been studied over a long period of time.

$$H_2(g) + I_2(g) \rightleftharpoons 2HI(g) \tag{1.27}$$

This reaction is found to be first-order in both H_2 and I_2. Therefore, the transition state (or activated complex) consists of one molecule of each. For many years, it was believed that the transition state had a structure like

$$\begin{array}{c} H ----- H \\ / \qquad\quad \backslash \\ I ------------ I \end{array}$$

However, more recent studies have shown that the I_2 molecules may be dissociated, and the activated complex may have a structure like

$$\begin{array}{c} H ----- H \\ / \qquad\quad \backslash \\ I \qquad\qquad I \end{array}$$

The rate law would still show a first-order dependence on I_2 because the molecules dissociate to produce two I atoms.

$$I_2(g) \rightleftharpoons 2I\cdot(g)$$

Therefore, the concentration of $I\cdot$ depends on the concentration of I_2 so that the reaction shows a first-order dependence on I_2. The reaction follows a rate law that is first-order in both H_2 and I_2, but the nature of the transition state was misunderstood for many years. Consequently, a reaction that was used as a model in numerous chemistry texts was described incorrectly.

1.5.2 Chain Mechanisms

The reaction of $H_2(g)$ with $Cl_2(g)$ follows the equation

$$H_2(g) + Cl_2(g) \longrightarrow 2HCl(g) \tag{1.28}$$

This equation looks as simple as the one representing the reaction of hydrogen with iodine. However, the reaction follows a completely different pathway. In this case, the reaction can be initiated by light (which has energy $E = h\nu$). In fact, a mixture of Cl_2 and H_2 will explode if a flashbulb is fired next to a *plastic* container holding a mixture of the two gases. The light causes some of the Cl_2 molecules to dissociate.

$$Cl_2(g) \xrightarrow{\text{h}\nu} 2Cl\cdot \tag{1.29}$$

We know that it is the Cl—Cl bond that is ruptured in this case since it is much weaker than the H—H bond (243 versus 435 kJ/mol). The next step in the process involves the reaction of Cl· and H_2.

$$Cl\cdot + H_2 \longrightarrow [Cl\cdots H\cdots H] \longrightarrow H\cdot + HCl \tag{1.30}$$

Then,

$$H\cdot + Cl_2 \longrightarrow [H\cdots Cl\cdots Cl] \longrightarrow Cl\cdot + HCl \tag{1.31}$$

This process continues with each step generating a radical, which can carry on the reaction in another step. Eventually, reactions such as

$$Cl\cdot + H\cdot \longrightarrow HCl \tag{1.32}$$

$$Cl\cdot + Cl\cdot \longrightarrow Cl_2 \tag{1.33}$$

$$H\cdot + H\cdot \longrightarrow H_2 \tag{1.34}$$

consume radicals without forming any new ones to cause the reaction to continue. The initial formation of Cl· as shown in Eq. (1.29) is called the *initiation* step, and the steps that form HCl and another radical are called *propagation* steps. The steps that cause the radicals to be consumed are called the *termination* steps. The entire process is often referred to as a *chain* or *free-radical* mechanism, and the rate law for this multistep process is very complicated. While the equation for the reaction looks as simple as that for the reaction of H_2 with I_2, the rate laws for the two reactions are quite different! The reaction of H_2 with Br_2 and the reaction of Cl_2 with hydrocarbons (as well as many other reactions of organic compounds) follow chain mechanisms. Likewise, the reaction between O_2 and H_2 follows a chain mechanism (see Chapter 4).

1.5.3 Substitution Reactions

Substitution reactions, which occur in all areas of chemistry, are those in which an atom or group of atoms is substituted for another. In fact, the reaction

$$A : B + : B' \longrightarrow A : B' + : B \qquad (1.35)$$

is an example of a Lewis acid-base reaction in which a stronger Lewis base, B', displaces a weaker one. Because Lewis bases are *nucleophiles*, this reaction is an example of a *nucleophilic substitution*. A typical reaction of this type is the reaction of tertiary butyl bromide, $t - (CH_3)_3CBr$, with hydroxide ion,

$$(1.36)$$

We could imagine this reaction as taking place in two different ways.

Case I. In this process, we will assume that Br^- leaves *before* the OH^- attaches. This process can be shown as

$$(1.37)$$

where [ ]* denotes the transition state, which in this case contains two ions. One of these contains a carbon atom having a positive charge, a species referred to as a *carbonium* ion or a *carbocation*. In this case, the transition state involves only one molecule of $t - (CH_3)_3CBr$, and the rate law is

$$\text{Rate} = k[t - (CH_3)_3CBr] \qquad (1.38)$$

If the reaction takes place by this pathway, it will be independent of OH⁻ concentration and follow the rate law shown in Eq. (1.38).

Case II. A second possible pathway for this reaction is one in which the OH⁻ starts to bond before the Br⁻ has completely left the $t - (CH_3)_3CBr$ molecule. In this process, the slow step involves both $t - (CH_3)_3CBr$ and OH⁻.

$$(1.39)$$

In this case, the formation of the transition state requires a molecule of $t - (CH_3)_3CBr$ and an OH⁻ ion in the rate-determining step, so the rate law is

$$\text{Rate} = k[t - (CH_3)_3CBr][OH^-] \tag{1.40}$$

When the reaction

$$t - (CH_3)_3CBr + OH^- \longrightarrow t - (CH_3)_3COH + Br^-$$

is studied in basic solutions, the rate is found to be independent of [OH⁻] concentration. Therefore, under these conditions, the reaction occurs by the pathway shown in Case I. This process is referred to as a *dissociative* pathway because it depends on the dissociation of the C—Br bond. Since the reaction is a nucleophilic substitution, and it is first-order, it is also labelled as an S_N1 process.

The fact that the reaction is first-order indicates that the slow step involves only a $t - (CH_3)_3CBr$ molecule. The second step, the addition of OH⁻ to the $t - (CH_3)_3C^+$, is fast under these conditions. At low concentrations of OH⁻, the second process may not be fast compared to the first. The reason for this is found in the Boltzmann Distribution Law. The transition state represents a high-energy state populated by a Boltzmann distribution. If a transition state were to be 50 kJ in energy above the reactant state, the relative populations at 300 K would be

$$\frac{n_2}{n_1} = e^{-E/RT} = e^{-50,000/8.3144 \times 300} = 2.0 \times 10^{-9}$$

Therefore, if the reactants represent a 1.0 **M** concentration, the transition state would be present at a concentration of 2.0×10^{-9} **M**. In a basic solution having a pH of 12.3, the $[OH^-]$ is 2×10^{-2} so there will be about 10^7 OH^- ions for every $t-(CH_3)_3C^+$! There is no surprise that the second step is fast when there is such an enormous excess of OH^- ions. On the other hand, at a pH of 5.0, the $[OH^-]$ is 10^{-9} **M**, and the kinetics of the reaction is decidedly different. Under these conditions, the second step is no longer very fast compared to the first, and the rate law now depends on the $[OH^-]$ as well. At low OH^- concentrations, the reaction follows a second-order rate law.

Since the reaction described involves *two* reacting species, there must be *some* conditions under which the reaction is *second*-order. The reason it appears as first-order at all is because of the large concentration of OH^- compared to the concentration of carbocation in the transition state. The reaction is in reality a *pseudo first-order* reaction.

Another interesting facet of this reaction is revealed by examining the transition state in the first-order process. In that case, the transition state consists of two *ions*. Because the reaction is being carried out in aqueous solution, these ions will be strongly solvated because of ion-dipole forces. Therefore, part of the energy required to break the C—Br bond will be recovered from the solvation enthalpies of the ions. It is generally true that the formation of a transition state in which charges must be separated is favored by carrying out the reaction in a polar solvent which will solvate the transition state species.

If the reaction

$$t-(CH_3)_3CBr + OH^- \longrightarrow t-(CH_3)_3COH + Br^-$$

is carried out in a solvent such as methyl alcohol, CH_3OH, it follows a second-order rate law. In this case, the CH_3OH is not as effective at solvating ions as is H_2O (because of size) so that the charges do not separate and the transition state is

$$OH \cdots \underset{\underset{CH_3}{|}}{\overset{\overset{\displaystyle H_3C \diagdown \diagup CH_3}{}}{C}} \cdots Br$$

If the reaction is carried out in a suitable mixture of CH_3OH and H_2O, the rate law is

$$\text{Rate} = k_1[t-(CH_3)_3CBr] + k_2[t-(CH_3)_3CBr][OH^-] \qquad (1.41)$$

indicating that both S_N1 (dissociative) and S_N2 (associative) pathways are being followed.

1.6 CATALYSIS

If there is a topic that is important to all branches of chemistry, it is catalysis. The gasoline used as a fuel, the polymers used in fabrics, the sulfuric acid used in an enormous range of chemical processes, and the ammonia used as a fertilizer are all produced by catalyzed reactions. In addition, many biological reactions are catalyzed by materials known as enzymes. It would be hard to overemphasize the importance of catalysis. In this section, we will describe some processes in which catalysts play an important role.

One of the important processes in organic chemistry is the reaction in which an alkyl group is attached to benzene. This reaction, known as the Friedel-Crafts reaction, can be shown as

$$RCl \quad + \quad \text{(benzene)} \quad \xrightarrow{\text{Catalyst}} \quad \text{(benzene-R)} \quad + \quad HCl \qquad (1.42)$$

where R is an alkyl group (CH_3, C_2H_5, etc.). The catalyst normally used in this reaction is $AlCl_3$, although other catalysts can be used. This reaction involves the interaction between the $AlCl_3$ and RCl to produce R^+,

$$AlCl_3 + RCl \rightleftharpoons AlCl_4^- + R^+ \qquad (1.43)$$

which occurs because $AlCl_3$ is a strong Lewis acid. Therefore, it interacts with an unshared electron pair on the Cl in the RCl molecule to cause it to be removed from the alkyl group. The R^+ then attacks the benzene ring to yield the final product, C_6H_5R. The function of the *acid* catalyst is to produce a *positive* species, which then attacks the other reactant.

Another reaction of this type is that in which an NO_2 group is introduced into an organic molecule. For example,

$$HNO_3 \quad + \quad \text{(benzene)} \quad \xrightarrow{H_2SO_4} \quad \text{(benzene-}NO_2\text{)} \quad + \quad H_2O \qquad (1.44)$$

In this case, the function of the H_2SO_4 is to protonate some of the HNO_3, which in turn leads to some NO_2^+ being formed.

$$HNO_3 + H_2SO_4 \rightleftharpoons HSO_4^- + H_2NO_3^+ \longrightarrow NO_2^+ + H_2O \qquad (1.45)$$

The NO_2^+, known as the *nitronium* ion, attacks the benzene ring to form the product, nitrobenzene.

$$NO_2^+ \ + \ \bighexagon \longrightarrow \overset{NO_2}{\bighexagon} \ + \ H^+ \qquad (1.46)$$

The *acid* catalyst, H_2SO_4, has functioned to generate a *positive* attacking species, which is generally the function of an acid catalyst. While we will not show examples here, it is the function of a *base* catalyst to generate some *negative* attacking species.

Hydrogenation reactions are reactions in which hydrogen is added to some compound, particularly unsaturated organic compounds. A large number of reactions of this type are of commercial importance, and all of them are catalyzed either by a solid catalyst (heterogeneous catalysis) or some catalyst in solution (homogeneous catalysis). One of the simplest reactions of this type is the hydrogenation of ethylene to produce ethane.

$$\underset{H}{\overset{H}{\diagdown}}C=C\underset{H}{\overset{H}{\diagup}} \ + \ H_2 \ \xrightarrow{\text{Catalyst}} \ H-\underset{H}{\overset{H}{\underset{|}{\overset{|}{C}}}}-\underset{H}{\overset{H}{\underset{|}{\overset{|}{C}}}}-H \qquad (1.47)$$

In this case, the catalyst is usually a metal such as platinum or nickel, and the function of the catalyst is of considerable interest. In order to understand how the catalyst works, it is necessary to know how hydrogen interacts with the metal.

We can picture a metal as being made up of spherical atoms in a close packing arrangement as shown in Figure 1.11. This figure also shows H and C_2H_4 adsorbed at active sites on the metal surface. In the process of adsorbing H_2 on the surface of the metal, some of the molecules become dissociated. Also, because metals that catalyze hydrogenation are those which form interstitial hydrides, some of the hydrogen penetrates to interstitial positions in the metal, which also favors the dissociation of H_2. Either of these processes produces some active hydrogen atoms, which can react with ethylene when it is also adsorbed on the metal surface. The details of the hydrogenation process are not completely understood, but the adsorption of H_2 and C_2H_4 is involved. Adsorption and dissolution of H_2 both favor the separation of the molecules, and the more reactive H atoms then react with the double bond in $H_2C=CH_2$, which subsequently leaves the surface of the metal as a molecule of C_2H_6.

Many reactions that are catalyzed by some solid in a process that is heterogeneous have as the essential step the adsorption of the reactants on the solid

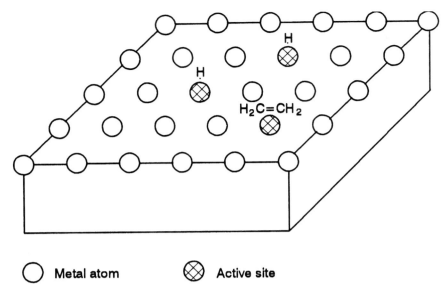

○ **Metal atom**   ⊗ **Active site**

Figure 1.11   A representation of the surface of a metal catalyst.

surface. The preparation of catalysts having surface characteristics that make them more effective in this type of interaction is currently a very important area of chemistry. In the cracking of hydrocarbons,

$$\text{RCH}_2\text{CH}_2\text{R}' \xrightarrow[500°\text{C}, 2 \text{ atm}]{\text{Catalyst}} \text{RH} + \text{R}'\text{CH}=\text{CH}_2 \tag{1.48}$$

the catalyst is usually a mixture of SiO_2 and Al_2O_3 in the form of a finely divided amorphous gel. The surface area of this material may be as high as 500 m^2/g, and the active sites behave as Lewis acids. Eventually, the surface of the catalyst becomes partially covered with carbon, and it must be regenerated thermally. This process causes the loss of some surface area by the rounding and smoothing of particles as they attempt to form a smaller surface area to minimize the number of particles on the surface. This motion of particles to form a surface of smaller area is known as *sintering*. Developed in the 1930s, these processes produce some low molecular weight hydrocarbons such as ethylene, propylene, and butenes, which are also useful in the preparation of polymers (e.g., polyethylene, polypropylene, etc.).

Another important process involving hydrocarbons is reforming. This type of process involves restructuring molecules so that they function better for some use, such as motor fuels. The early catalysts for this type of process were

Al_2O_3 containing some Cr_2O_3 or Mo_2O_3, but a platinum catalyst is more widely used now. Typical reactions of this type might be

$$CH_3(CH_2)_4CH_3 \xrightarrow{\text{Catalyst}} C_6H_{12} + H_2 \qquad (1.49)$$

$$C_6H_{12} \xrightarrow{\text{Catalyst}} C_6H_6 + 3H_2 \qquad (1.50)$$

The benzene produced is used as a solvent, in the preparation of polystyrene, and in many other applications. While a comprehensive description of catalysis is beyond the scope of this chapter, it is, however, a topic of enormous importance in modern chemistry and it will be discussed in more detail in Chapter 4.

This chapter has provided a review of some of the topics covered in earlier courses and an introduction to several of the topics that will be treated in more detail in subsequent chapters. We will begin the more detailed study in the next chapter by considering the treatment of systems following more complicated rate laws.

REFERENCES FOR FURTHER READING

Alberty, R. A., Silber, R. J. (1992) *Physical Chemistry,* Wiley, New York, Chapter 19. An outstanding introduction to chemical kinetics.

Cox, B. G. (1994) *Modern Liquid Phase Kinetics,* Oxford, New York. The first three chapters of this book provide a good introduction to general kinetics.

Dence, J. B., Gray, H. B., Hammond, G. S. (1968) *Chemical Dynamics,* Benjamin, New York. A good survey of kinetic studies on many types of reactions, especially in solutions.

Laidler, K. J. (1987) *Chemical Kinetics,* 3d ed., Harper-Collins, New York. Latest edition of a classic text in chemical kinetics.

Moore, J. W., Pearson, R. G. (1981) *Kinetics and Mechanism,* 3d ed., Wiley, New York. One of the standard books on chemical kinetics.

PROBLEMS

1. For the reaction A ⟶ products, the following data were obtained.

Time (hrs)	[A], **M**	Time (hrs)	[A], **M**
0	1.24	6	0.442
1	0.960	7	0.402
2	0.775	8	0.365
3	0.655	9	0.335
4	0.560	10	0.310
5	0.502		

(a) Make appropriate plots of these data to test them for fitting zero- , first-, and second-order rate laws. Test all three even if you happen to guess the correct rate law on the first trial. (b) Determine the rate constant for the reaction. (c) Using the rate law that you have determined, calculate the half-life for the reaction. (d) At what time will the concentration of A be 0.380 **M**?

2. For the reaction X ⟶ Y, the following data were obtained.

Time (min)	[X], **M**	Time (min)	[X], **M**
0	0.500	60	0.240
10	0.443	70	0.212
20	0.395	80	0.190
30	0.348	90	0.171
40	0.310	100	0.164
50	0.274		

(a) Make appropriate plots of these data to determine the reaction order. (b) Determine the rate constant for the reaction. (c) Using the rate law you have determined, calculate the half-life for the reaction. (d) Calculate how long it will take for the concentration of X to be 0.330 **M**.

3. If the half-life for the reaction

$$C_2H_5Cl \longrightarrow C_2H_4 + HCl$$

is the same when the initial concentration of C_2H_5Cl is 0.0050 **M** and 0.0078 **M**, what is the rate law?

4. When the reaction A + 2B ⟶ D is studied kinetically, it is found that the rate law is R = k[A][B]. Propose a mechanism that is consistent with this observation. Explain how the proposed mechanism is consistent with the rate law.

5. The decomposition of A to produce B can be written as A ⟶ B. (a) When the initial concentration of A is 0.012 **M**, the rate is 0.0018 **M** min⁻¹,

and when the initial concentration of A is 0.024 **M**, the rate is 0.0036 **M** min^{-1}. Write the rate law for the reaction. (b) If the activation energy for the reaction is 268 kJ mol^{-1} and the rate constant at 660 K is 8.1 x 10^{-3} sec^{-1}, what will be the rate constant at 690 K?

6. For the reaction X $\rightleftharpoons$ Y, the following data were obtained for the forward (k_f) and reverse (k_r) reactions.

T, K	400	410	420	430	440
k_f, sec^{-1}	0.161	0.279	0.470	0.775	1.249
$10^3 \times k_r$, sec^{-1}	0.159	0.327	0.649	1.25	2.323

Use these data to determine the activation energy for the forward and reverse reactions. Draw a reaction energy profile for the reaction.

7. For a reaction A $\longrightarrow$ B, the following data were collected when a kinetic study was carried out at several temperatures.

[A], **M**

t, min	°C 25	30	35	40	45
0	0.750	0.750	0.750	0.750	0.750
15	0.648	0.622	0.590	0.556	0.520
30	0.562	0.530	0.490	0.440	0.400
45	0.514	0.467	0.410	0.365	0.324
60	0.460	0.410	0.365	0.315	0.270
75	0.414	0.378	0.315	0.275	0.235
90	0.385	0.336	0.290	0.243	0.205

(a) Use one of the data sets and make appropriate plots on a graph to determine the order of the reaction. (b) After you have determined the correct rate law, determine graphically the rate constant at each temperature. (c) Having determined the rate constants at several temperatures, determine the activation energy.

8. Suppose a solid metal catalyst has a surface area of 1000 cm^2. (a) If the distance between atomic centers is 145 pm and the structure of the metal is simple cubic, how many metal atoms are exposed on the surface? (b) Assuming that the number of active sites on the metal has an equilibrium concentration of adsorbed gas, A, and that the rate of the reaction A $\longrightarrow$ B is 1.00 × 10^{-6} moles/sec^{-1}, what fraction of the metal atoms on the surface have a molecule of A adsorbed? Assume each molecule is adsorbed for 0.1 sec before reacting.

9. A reaction has a rate constant of 0.264 **M**$^{-1}$ sec^{-1} at 45.6 °C and 0.833 **M**$^{-1}$ sec^{-1} at 58.8 °C. What is the activation energy for the reaction?

10. The rate constant for a reaction is 0.322 min^{-1} at 33.0 °C, and the activation energy is 58.8 kJ mol^{-1}. What will be the rate constant at 70 °C?

11. What is the activation energy for a reaction whose rate doubles when the temperature is raised from 25 °C to 40 °C? What is the half-life of this reaction at 35 °C?
12. When initially present at 1.00 **M** concentration, one-tenth of a sample reacts in 36 minutes. (a) What is the half-life of the material if the reaction is first order? (b) What is the half-life if the reaction is second-order?
13. Strontium-83 has a half-life of 32.4 hours. If you receive a sample of pure ^{89}Sr and must complete a study of the nuclide before 3.00% of the material decays, how long do you have to complete the required study?

chapter 2

KINETICS OF MORE COMPLEX SYSTEMS

In Chapter 1, we showed the mathematical treatment of simple systems in which the rate law is a function of the concentration of only one reactant. However, a large proportion of reactions either involve more than one reactant in the rate-determining step or give more than one product. The description of such systems is necessary for a survey of chemical kinetics and we will now treat some of these cases.

2.1 SECOND-ORDER CASE, FIRST-ORDER IN TWO COMPONENTS

A reaction such as

$$A + B \longrightarrow \text{Products} \qquad (2.1)$$

which follows the rate law

$$-\frac{d[A]}{dt} = -\frac{d[B]}{dt} = k[A]^1 [B]^1 \qquad (2.2)$$

is second-order overall. Therefore, such a reaction is referred to as a second-order mixed case since two reactants are involved. While the second-order case in one component was considered in Chapter 1, the second-order mixed case is somewhat more complex. However, this type of rate law occurs very often because two component reactions are very numerous. In Eq. (2.1), the balancing coefficients are equal so the amount of A reacting, $[A]_o - [A]$, must be equal to the amount of B reacting, $[B]_o - [B]$. Therefore, we can write

$$[A]_o - [A] = [B]_o - [B] \qquad (2.3)$$

or

$$[B] = [B]_o - [A]_o + [A] \tag{2.4}$$

Therefore, substituting for [B] in Eq. (2.2) yields

$$-\frac{d[A]}{dt} = k[A]([B]_o - [A]_o + [A]) \tag{2.5}$$

which can be written as

$$-\frac{d[A]}{[A]([B]_o - [A]_o + [A])} = k\,dt \tag{2.6}$$

The fraction can be separated as follows.

$$\frac{1}{[A]([B]_o - [A]_o + [A])} = \frac{C_1}{[A]} + \frac{C_2}{[B]_o - [A]_o + [A]} \tag{2.7}$$

where C_1 and C_2 are constants to be determined. However,

$$\frac{C_1}{[A]} + \frac{C_2}{[B]_o - [A]_o + [A]} = \frac{C_1([B]_o - [A]_o + [A]) + C_2[A]}{[A]([B]_o - [A]_o + [A])} \tag{2.8}$$

Because the two sides of the equation are equal and because they are equal to the left-hand side of Eq. (2.7), it is clear that

$$C_1([B]_o - [A]_o + [A]) + C_2[A] = 1$$

or

$$C_1[B]_o - C_1[A]_o + C_1[A] + C_2[A] = 1 \tag{2.9}$$

After an infinitely long time, [A] = 0, so

$$C_1[B]_o - C_1[A]_o = 1 = C_1([B]_o - [A]_o) \tag{2.10}$$

Therefore, by combining Eq. (2.10) with Eq. (2.9), it follows that after an infinite time of reaction,

$$C_1[B]_o - C_1[A]_o + C_1[A] + C_2[A] = 1 = C_1[B]_o - C_1[A]_o$$

or

$$C_1[A] + C_2[A] = 0 \tag{2.11}$$

Since [A] is not zero except after an infinite time, it follows that

$$C_1 + C_2 = 0$$

or

$$C_1 = -C_2 \qquad (2.12)$$

Therefore,

$$1 = C_1([B]_o - [A]_o)$$

so that

$$C_1 = \frac{1}{[B]_o - [A]_o} \quad \text{and} \quad C_2 = -\frac{1}{[B]_o - [A]_o} \qquad (2.13)$$

Eq. (2.6) can now be written as

$$-\frac{d[A]}{([B]_o - [A]_o)[A]} + \frac{d[A]}{([B]_o - [A]_o)\,([B]_o - [A]_o + [A])} = k\,dt \qquad (2.14)$$

which can be written as

$$\frac{1}{[B]_o - [A]_o}\left(-\frac{d[A]}{[A]}\right) + \left(\frac{1}{[B]_o - [A]_o}\right)\frac{d[A]}{[B]_o - [A]_o + [A]} = k\,dt$$

Since $1/([B]_o - [A]_o)$ is a constant, upon integration the result is

$$\frac{1}{[B]_o - [A]_o}\ln\frac{[A]_o}{[A]} + \frac{1}{[B]_o - [A]_o}\ln\frac{([B]_o - [A]_o + [A])}{[B]_o} = kt \qquad (2.15)$$

Combining the two terms on the left-hand side of Eq. (2.15), we obtain

$$\frac{1}{[B]_o - [A]_o}\ln\frac{[A]_o\,([B]_o - [A]_o + [A])}{[A][B]_o} = kt \qquad (2.16)$$

However,

$$[B]_o - [A]_o + [A] = [B]$$

By substituting this result in Eq. (2.16), we obtain

$$\frac{1}{[B]_o - [A]_o}\ln\frac{[A]_o[B]}{[A][B]_o} = kt \qquad (2.17)$$

By separating the logarithm factors, we can write

$$\frac{1}{[B]_o - [A]_o} \ln \frac{[A]_o}{[B]_o} \frac{[B]}{[A]} = \frac{1}{[B]_o - [A]_o} \left(\ln \frac{[A]_o}{[B]_o} + \ln \frac{[B]}{[A]} \right) = kt$$

This equation can be simplified to yield

$$\frac{1}{[B]_o - [A]_o} \ln \frac{[A]_o}{[B]_o} + \frac{1}{[B]_o - [A]_o} \ln \frac{[B]}{[A]} = kt \qquad \textbf{(2.18)}$$

or

$$\ln \frac{[A]_o}{[B]_o} + \ln \frac{[B]}{[A]} = kt([B]_o - [A]_o) \qquad \textbf{(2.19)}$$

Therefore, a plot of $\ln([B]/[A])$ versus t is linear with a slope of $k([B]_o - [A]_o)$ and an intercept of $\ln([B]_o/[A]_o)$.

Frequently, an alternate way of describing second-order processes of this type is employed in which the extent of reaction, x, is used. If the reaction is one in which the balancing coefficients are both 1, the amounts of A and B used are equal. Therefore, we can write

$$\frac{dx}{dt} = k(a - x)(b - x) \qquad \textbf{(2.20)}$$

where a and b are the initial concentrations of A and B, respectively. It follows that $(a - x)$ and $(b - x)$ are the concentrations of A and B remaining after some extent of reaction. Integration of this equation using the method of integration by parts yields

$$\frac{1}{b - a} \ln \frac{a(b - x)}{b(a - x)} = kt \qquad \textbf{(2.21)}$$

Another form of this equation is

$$\frac{1}{a - b} \ln \frac{b(a - x)}{a(b - x)} = kt \qquad \textbf{(2.22)}$$

The hydrolysis of an ester can be used to illustrate the type of second-order process described in this section. The hydrolysis of ethyl acetate produces ethyl alcohol and acetic acid.

$$CH_3COOC_2H_5 + H_2O \longrightarrow CH_3COOH + C_2H_5OH \qquad \textbf{(2.23)}$$

When this reaction is carried out in basic solution, OH^- is consumed by the reaction

$$CH_3COOH + OH^- \longrightarrow CH_3COO^- + H_2O \qquad \textbf{(2.24)}$$

Table 2.1   Analysis of kinetic data for hydrolysis of ethyl acetate in NaOH solution.

Time (min)	ml NaOH for back-titrating	Concentration of NaOH reacted	$\dfrac{(a-x)}{(b-x)}$	$\ln\dfrac{(a-x)}{(b-x)}$
1	13.83	0.00076	1.83	0.604
3	14.60	0.00112	1.90	0.642
5	15.40	0.00200	2.10	0.742
10	16.60	0.00300	2.50	0.916
20	17.95	0.00404	3.38	1.22
35	19.02	0.00488	5.57	1.72
55	19.68	0.00544	12.70	2.54
75	19.85	0.00558	25.27	3.23

In the experiment described here, 125 ml of $CH_3COOC_2H_5$ at 30°C was mixed with 125 ml of NaOH solution so that the concentrations were 0.00580 **M** and 0.0100 **M**, respectively. If an aliquot is removed from the reaction vessel and quenched by putting it in a solution containing an excess but known amount of HCl, the remaining OH⁻ will quickly be neutralized. By back-titrating the excess HCl, one can determine the amount of HCl that is left unreacted, which makes it possible to determine how much OH⁻ has been consumed.

The data shown in Table 2.1 were obtained for the hydrolysis of 0.00580 **M** $CH_3COOC_2H_5$ at 30°C in 0.0100 **M** NaOH. The HCl used for quenching the reaction was 0.0203 **M**, and the titration of the excess HCl was carried out using 0.0200 **M** NaOH. Each aliquot contained 25 ml of the reaction mixture.

When the data shown in Table 2.1 were used to prepare a second-order plot for the hydrolysis reaction, the results are as shown in Figure 2.1. As expected, the plot is linear and the slope is calculated as $(3.02 - 0.92)/(70 - 10) = 0.0350$ min⁻¹. The rate constant is the slope/$(a - b)$. In this case, $(a - b)$ is $(0.0100 - 0.00580) = 0.00420$, which leads to $k = 8.33$ **M**⁻¹ min⁻¹ or 0.139 **M**⁻¹ sec⁻¹. As mentioned earlier, reactions following this type of rate law are numerous.

2.2   OTHER REACTION ORDERS

If a reaction takes place for which only one reactant is involved, the general rate law can be written as

$$-\frac{d[A]}{dt} = k[A]^n \tag{2.25}$$

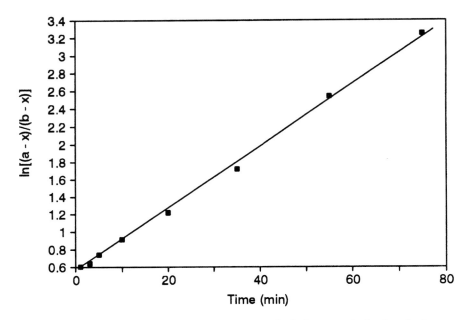

Figure 2.1 Second-order plot for the hydrolysis of ethyl acetate in basic solution.

When n is not equal to 1, integration of this equation gives

$$\frac{1}{[A]^{n-1}} - \frac{1}{[A]_o^{n-1}} = (n-1)kt \tag{2.26}$$

It is easy to show that the half-life can be written as

$$t_{1/2} = \frac{2^{n-1} - 1}{(n-1)k[A]_o^{n-1}} \tag{2.27}$$

In this case, n may be either a fraction or an integer.

Earlier, we worked through the details of the second-order reaction which is first-order in two components. We consider briefly here the various third-order cases (which are worked out in detail by Benson (1960)). The simplest case, that of

$$-\frac{d[A]}{dt} = k[A]^3 \tag{2.28}$$

is easily handled to give

$$\frac{1}{[A]^2} - \frac{1}{[A]_o^2} = 2kt \tag{2.29}$$

from which we find

$$t_{1/2} = \frac{3}{2k[A]_o^2} \tag{2.30}$$

Also, a plot of $1/[A]^2$ versus time should be linear with a slope of $2k$.

A third-order reaction can also arise from a reaction such as

$$aA + bB \longrightarrow \text{Products} \tag{2.31}$$

and

$$-\frac{d[A]}{dt} = k[A]^2[B] \tag{2.32}$$

If the stoichiometry is such that $[A]_o - [A] = [B]_o - [B]$, the integrated rate law is

$$\frac{1}{[B]_o - [A]_o}\left(\frac{1}{[A]} - \frac{1}{[A]_o}\right) + \frac{1}{([B]_o - [A]_o)^2} \ln \frac{[B]_o[A]}{[A]_o[B]} = kt \tag{2.33}$$

However, if the stoichiometry is such that

$$[A]_o - [A] = 2([B]_o - [B])$$

then the integrated rate law is

$$\frac{2}{(2[B]_o - [A]_o)}\left(\frac{1}{[A]} - \frac{1}{[A]_o}\right) + \frac{2}{(2[B]_o - [A]_o)^2} \ln \frac{[B]_o[A]}{[A]_o[B]} = kt \tag{2.34}$$

A reaction such as

$$A + B + C \longrightarrow \text{Products} \tag{2.35}$$

where the stoichiometry is such that

$$[A]_o - [A] = [B]_o - [B] = [C]_o - [C]$$

can follow a rate law such as

$$-\frac{d[A]}{dt} = k[A][B][C] \tag{2.36}$$

After making the substitutions $[B] = [B]_o - [A]_o + [A]$ and $[C] = [C]_o - [A]_o + [A]$, some very laborious mathematics yields the integrated form

$$\frac{1}{LMN} \ln \left(\frac{[A]}{[A]_o}\right)^M \left(\frac{[B]}{[B]_o}\right)^N \left(\frac{[C]}{[C]_o}\right)^L = kt \tag{2.37}$$

where $L = [A]_o - [B]_o$, $M = [B]_o - [C]_o$, and $N = [C]_o - [A]_o$. Space will not be devoted in this brief book to the mathematical details of following other orders systems although many have been described.

2.3   PARALLEL FIRST-ORDER REACTIONS

Suppose a compound, A, undergoes reaction by forming several products, B, C, and D, at different rates. We can show this system as

$$A \xrightarrow{k_1} B$$

$$A \xrightarrow{k_2} C \tag{2.38}$$

$$A \xrightarrow{k_3} D$$

The disappearance of A is the sum of the three rates and

$$-\frac{d[A]}{dt} = k_1[A] + k_2[A] + k_3[A] = (k_1 + k_2 + k_3)[A] \tag{2.39}$$

Then

$$-\int_{[A]_o}^{[A]} \frac{d[A]}{[A]} = (k_1 + k_2 + k_3)\int_0^t dt \tag{2.40}$$

Integration yields

$$\ln \frac{[A]_o}{[A]} = (k_1 + k_2 + k_3)t \tag{2.41}$$

Equation (2.41) can also be written as

$$[A] = [A]_o \exp(-(k_1 + k_2 + k_3)t) \tag{2.42}$$

The concentration of B changes according to the equation

$$\frac{d[B]}{dt} = k_1[A] = k_1[A]_o \exp(-(k_1 + k_2 + k_3)t) \tag{2.43}$$

Letting $k = k_1 + k_2 + k_3$ and rearranging yields

$$d[B] = k_1[A]_o e^{-kt} \, dt \tag{2.44}$$

so that

$$\int_{[B]_o}^{[B]} d[B] = k_1[A]_o \int_0^t e^{-kt} \, dt \tag{2.45}$$

Integration of Eq. (2.45) between appropriate limits leads to

$$[B] = [B]_o + \frac{k_1[A]_o}{k}(1 - e^{-kt})$$

Substituting for k gives

$$[B] = [B]_o + \frac{k_1[A]_o}{k_1 + k_2 + k_3}(1 - \exp -(k_1 + k_2 + k_3)t) \tag{2.46}$$

In a similar way, the concentrations of C and D as functions of time are obtained.

$$[C] = [C]_o + \frac{k_2[A]_o}{k_1 + k_2 + k_3}(1 - \exp -(k_1 + k_2 + k_3)t) \tag{2.47}$$

$$[D] = [D]_o + \frac{k_3[A]_o}{k_1 + k_2 + k_3}(1 - \exp -(k_1 + k_2 + k_3)t) \tag{2.48}$$

If, as is the usual case, $[B]_o = [C]_o = [D]_o = 0$, then the ratio of Eqs. (2.46) and (2.47) gives

$$\frac{[B]}{[C]} = \frac{\dfrac{k_1[A]_o}{k_1 + k_2 + k_3}(1 - \exp -(k_1 + k_2 + k_3)t)}{\dfrac{k_2[A]_o}{k_1 + k_2 + k_3}(1 - \exp -(k_1 + k_2 + k_3)t)} = \frac{k_1}{k_2} \tag{2.49}$$

Similarly, it is easy to show that

$$\frac{[B]}{[D]} = \frac{k_1}{k_3} \tag{2.50}$$

The case where $[A]_o = 1.00$ **M** and $k_1 = 0.03$, $k_2 = 0.02$, and $k_3 = 0.01$ min^{-1} is shown graphically in Figure 2.2. In this diagram, the sum of $[A] + [B] + [C] + [D]$ = 1.00 **M**. For all time values, the ratios of concentrations $[B]:[C]:[D]$ is the same as $k_1 : k_2 : k_3$.

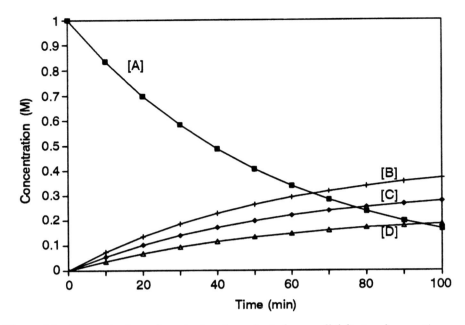

Figure 2.2 Concentration of reactant and products for parallel first-order reactions.

2.4 SERIES FIRST-ORDER REACTIONS

It is by no means uncommon for a chemical reaction to take place in steps.

$$A \xrightarrow{k_1} B \xrightarrow{k_2} C \qquad (2.51)$$

In this case, B is known as an *intermediate* because it is not the final product. A similar situation is very common in nuclear chemistry where a nuclide decays to a daughter, which is also radioactive and undergoes decay. For simplicity, only the case of first-order reactions will be treated here.

The rate of disappearance of A can be written as

$$-\frac{d[A]}{dt} = k_1[A] \qquad (2.52)$$

The change in concentration of B with time is

$$\frac{d[B]}{dt} = k_1[A] - k_2[B] \qquad (2.53)$$

where the term $k_1[A]$ represents the formation of B from A, and the term $-k_2[B]$ represents the reaction of B to form C. The rate of formation of C can be expressed as

$$\frac{d[C]}{dt} = k_2[B] \tag{2.54}$$

If the stoichiometry as shown in Eq. (2.51) is followed, it is obvious that

$$[A] + [B] + [C] = [A]_o \tag{2.55}$$

Eq. (2.52) can be integrated immediately to give

$$[A] = [A]_o \exp(-k_1 t) \tag{2.56}$$

Substituting for [A] in Eq. (2.53) gives

$$\frac{d[B]}{dt} = k_1[A]_o \exp(-k_1 t) - k_2[B] \tag{2.57}$$

which can be written as

$$\frac{d[B]}{dt} + k_2[B] - k_1[A]_o \exp(-k_1 t) = 0 \tag{2.58}$$

This is a linear differential equation with constant coefficients. If we assume a solution of the form

$$[B] = u \exp(-k_2 t)$$

then

$$\frac{d[B]}{dt} = -uk_2 \exp(-k_2 t) + \exp(-k_2 t)\frac{du}{dt}$$

Substituting the right-hand side of this equation for d[B]/dt in Eq. (2.58), we obtain

$$-uk_2 \exp(-k_2 t) + \exp(-k_2 t)\frac{du}{dt} = k_1[A]_o \exp(-k_1 t) - uk_2 \exp(-k_2 t)$$

which simplifies to

$$\exp(-k_2 t)\frac{du}{dt} = k_1[A]_o \exp(-k_1 t)$$

Dividing both sides of this equation by $\exp(-k_2 t)$ gives

$$\frac{du}{dt} = k_1[A]_o \exp(-(k_1 - k_2)t) \tag{2.59}$$

Integration of this equation yields

$$u = \frac{k_1}{k_2 - k_1}[A]_\circ \exp(-(k_1 - k_2)t) + C \tag{2.60}$$

where C is a constant. Having assumed a solution of the form

$$[B] = u \exp\left(-k_2 t\right)$$

we obtain

$$[B] = u \exp(-k_2 t) = \frac{k_1[A]_\circ}{k_2 - k_1}\exp(-k_1 t) + C \cdot \exp(-k_2 t) \tag{2.61}$$

If we let $[B]_\circ$ be the concentration of B present at t = 0, Eq. (2.61) reduces to

$$[B]_\circ = \frac{k_1[A]_\circ}{k_2 - k_1} + C$$

so that the final equation for the concentration of B is

$$[B] = \frac{k_1[A]_\circ}{k_2 - k_1}(\exp(-k_1 t) - \exp(-k_2 t)) + [B]_\circ \exp(-k_2 t) \tag{2.62}$$

The first term on the right-hand side of Eq. (2.62) gives the reaction of B produced by disappearance of A, while the second term gives the reaction of any B initially present. If, as is usual, $[B]_\circ = 0$, Eq. (2.62) reduces to

$$[B] = \frac{k_1[A]_\circ}{k_2 - k_1}(\exp(-k_1 t) - \exp(-k_2 t)) \tag{2.63}$$

If this result and that shown for [A] in Eq. (2.56) are substituted into Eq. (2.55), we obtain

$$[C] = [A]_\circ\left[1 - \frac{1}{k_2 - k_1}(k_2 \exp(-k_1 t) - k_1 \exp(-k_2 t))\right] \tag{2.64}$$

A number of interesting cases can arise depending on the relative magnitudes of k_1 and k_2. Figure 2.3 shows the unlikely case where $k_1 = 2k_2$. This case is unlikely because the intermediate, B, is usually more reactive than A, as shown in Figure 2.4 where $k_2 = 2k_1$. In this case, it is apparent that there is a less rapid decrease in [A] and a slower buildup of B in the system. Because of the particular relationship chosen for the rate constants ($k_1 = 2k_2$ and $k_2 = 2k_1$), production of C is unchanged in the two cases. Figure 2.5 shows the case where $k_2 = 10k_1$, as

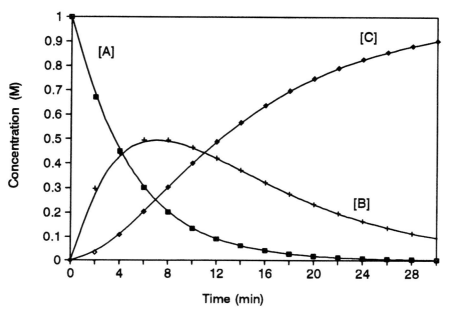

Figure 2.3   Series first-order reactions where $[A]_o = 1.00$ **M**, $k_1 = 0.200$ and $k_2 = 0.100$ min⁻¹.

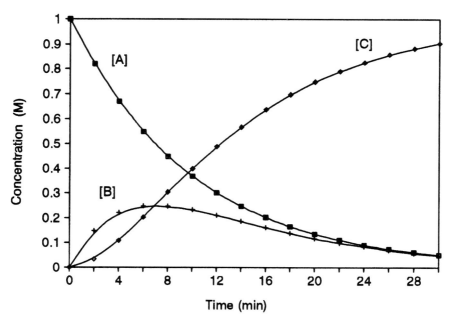

Figure 2.4   Series first-order reactions where $[A]_o = 1.00$ **M**, $k_1 = 0.100$ and $k_2 = 0.200$ min⁻¹.

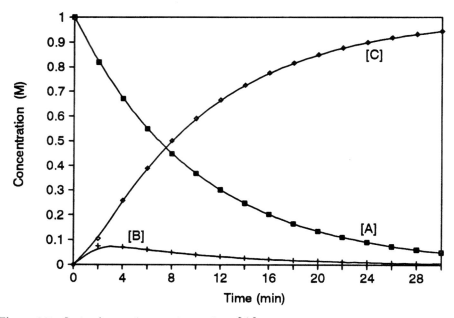

Figure 2.5   Series first-order reactions where $[A]_o = 1.00$ **M**, $k_1 = 0.100$ and $k_2 = 1.00$ min^{-1}.

a realistic example of a system with a reactive intermediate. In this case, the concentration of B is always low, which is more likely for an intermediate. Further, over a large extent of reaction, [B] is essentially a constant. When $k_2 > k_1$, there is a low and essentially constant concentration of intermediate, B. Therefore, d[B]/dt is essentially 0, which can be shown as follows. For this system of first-order reactions,

$$[A] + [B] + [C] = [A]_o$$

which, because [B] is nearly 0, is approximated by

$$[A] + [C] = [A]_o$$

Taking the derivatives with respect to time,

$$\frac{d[A]}{dt} + \frac{d[B]}{dt} + \frac{d[C]}{dt} = 0$$

and

$$\frac{d[A]}{dt} + \frac{d[C]}{dt} = 0$$

Therefore $d[B]/dt = 0$, and $[B]$ remains constant throughout most of the reaction. For the case where $k_2 = 10k_1$ and $[A]_o = 1.00$ **M** (shown in Figure 2.5), $[B]$ never goes above 0.076 **M** and it varies only from 0.076 to 0.033 **M** from t = 2 min to t = 12 min where $[A]$ varies from 0.819 to 0.301 **M** and $[C]$ varies from 0.105 **M** to 0.666 **M**. The approximation of considering the concentration of the intermediate to be essentially constant is called the *steady-state approximation*.

It is clear from Figures 2.3 through 2.5 that $[B]$ goes through a maximum as expected. The time necessary to reach that maximum concentration of B, t_m, can easily be calculated. At that time, $d[B]/dt = 0$. If Eq. (2.63) is differentiated with respect to time,

$$\frac{d[B]}{dt} = \frac{-k_1 k_1 [A]_o}{k_2 - k_1} \exp(-k_1 t) + \frac{k_1 k_2 [A]_o}{k_2 - k_1} \exp(-k_2 t) = 0 \tag{2.65}$$

Therefore,

$$\frac{k_1 k_1 [A]_o}{k_2 - k_1} \exp(-k_1 t) = \frac{k_1 k_2 [A]_o}{k_2 - k_1} \exp(-k_2 t)$$

Cancelling like terms from both sides of the equation gives

$$k_1 \exp(-k_1 t) = k_2 \exp(-k_2 t)$$

which can be written as

$$k_1 / k_2 = \exp(-k_2 t) / \exp(-k_1 t) = \exp(-k_2 t)(\exp(k_1 t)) = \exp((k_1 - k_2)t)$$

Taking the logarithm of both sides of the equation gives

$$\ln(k_1 / k_2) = \left(k_1 - k_2\right)t$$

which yields the time to reach the maximum in the $[B]$ curve,

$$t_m = \frac{\ln(k_1 / k_2)}{k_1 - k_2} \tag{2.66}$$

Although several other cases of series of reactions have been described mathematically, we will leave these discussions to more advanced books. The basic principles have been adequately demonstrated here.

2.5   REVERSIBLE REACTIONS

Many reactions do not proceed to completion, and the extent of reversibility must be considered from the early stages. The first-order case is the simplest.

$$A \underset{k_{-1}}{\overset{k_1}{\rightleftharpoons}} B \tag{2.67}$$

It is clear that

$$-\frac{d[A]}{dt} = k_1[A] - k_{-1}[B] \tag{2.68}$$

Assuming that only A is initially present,

$$[B] = [A]_o - [A] \tag{2.69}$$

Substituting for [B] in Eq. (2.68) gives

$$-\frac{d[A]}{dt} = k_1[A] - k_{-1}([A]_o - [A]) \tag{2.70}$$

Therefore,

$$-d[A] = \{k_1[A] - k_{-1}([A]_o - [A])\}dt$$

or

$$-\frac{d[A]}{(k_1 + k_{-1})[A] - k_{-1}[A]_o} = dt$$

This equation must be integrated between the limits of $[A]_o$ at $t = 0$ and $[A]$ at time t.

$$-\int_{[A]_o}^{[A]} \frac{d[A]}{(k_1 + k_{-1})[A] - k_{-1}[A]_o} = \int_0^t dt$$

This integral is of the form

$$\int \frac{dx}{a + bx} = \frac{1}{b} \ln(a + bx)$$

where $b = (k_1 + k_{-1})$ and $a = -k_{-1}[A]_o$. Then

$$-\frac{1}{k_1 + k_{-1}} \ln\{(k_1 + k_{-1})[A] - k_{-1}[A]_o\} \bigg|_{[A]_o}^{[A]} = t \tag{2.71}$$

Multiplying both sides of Eq. (2.71) by $(k_1 + k_{-1})$ and expanding by making use of the limits [A] and $[A]_o$, we obtain

$$-[\ln\{(k_1 + k_{-1})[A] - k_{-1}[A]_o\} - \ln\{(k_1 + k_{-1})[A]_o - k_{-1}[A]_o\}] = (k_1 + k_{-1})t$$

which simplifies to give the integrated rate equation

$$\ln \frac{k_1[A]_o}{(k_1 + k_{-1})[A] - k_{-1}[A]_o} = (k_1 + k_{-1})t \qquad (2.72)$$

As equilibrium is approached at infinite time, $t \longrightarrow t_f$ and $d[A]/dt = 0$.

$$\frac{d[A]}{dt} = 0 = k_1[A]_f - k_{-1}[B]_f$$

and

$$k_1[A]_f = k_{-1}[B]_f = k_{-1}([A]_o - [A]_f) = k_{-1}[A]_o - k_{-1}[A]_f$$

so

$$[A]_f = \frac{k_{-1}[A]_o}{k_1 + k_{-1}} \qquad (2.73)$$

and

$$[A]_o = ((k_1 + k_{-1})/k_{-1})[A]_f$$

Substituting this value for $[A]_o$ in Eq. (2.72),

$$\ln \frac{\dfrac{k_1(k_1 + k_{-1})[A]_f}{k_{-1}}}{(k_1 + k_{-1})[A] - k_{-1}[(k_1 + k_{-1})[A]_f / k_{-1}]} = (k_1 + k_{-1})t$$

or

$$\ln \frac{\dfrac{k_1(k_1 + k_{-1})[A]_f}{k_{-1}}}{(k_1 + k_{-1})([A] - [A]_f)} = (k_1 + k_{-1})t$$

Now $k_1[A]_f = k_{-1}([A]_o - [A]_f)$ so

$$\ln \frac{k_{-1}([A]_o - [A]_f)\dfrac{k_1 + k_{-1}}{k_{-1}}}{(k_1 + k_{-1})([A] - [A]_f)} = \ln \frac{(k_1 + k_{-1})([A]_o - [A]_f)}{(k_1 + k_{-1})([A] - [A]_f)} = (k_1 + k_{-1})t$$

Therefore, the equation for the infinite time or final condition is

$$\ln \frac{[A]_o - [A]_f}{[A] - [A]_f} = (k_1 + k_{-1})t \tag{2.74}$$

which can be written as

$$[A] = [A]_f + ([A]_o - [A]_f)\exp(-(k_1 + k_{-1})t) \tag{2.75}$$

For this reversible first-order system, a plot of $\ln([A] - [A]_f)$ versus time should be linear with a slope of $-(k_1 + k_{-1})$.

Let us consider the case where $[A]_o = 1.000$ M, $k_1 = 0.050$ min^{-1} and $k_{-1} = 0.010$ min^{-1} for which the equilibrium constant, K, is 5.00. Then, for the reaction A $\rightleftharpoons$ B,

$$K = \frac{[B]_f}{[A]_f} = 5.00 = \frac{x}{1.00 - x}$$

so that $x = 0.167$, which is the equilibrium concentration of A. The variation of [A] with time for this case is shown in Table 2.2.

These data were used to prepare the plot of $\ln([A] - [A]_f)$ versus time shown in Figure 2.6. As shown in Eq. (2.74), this linear plot has a slope that is $-(k_1 + k_{-1})$.

If the reaction is very slow, it is difficult to determine $[A]_f$ accurately and therefore, the limiting accuracy $[A]_f \pm \Delta[A]_f$ is actually known. Using the arbitrary error of 0.033 M in $[A]_f$ with the case described above, $[A]_f$ could vary from 0.133 M to 0.200 M. When these errors in $[A]_f$ are introduced and Eq. (2.75) is used to calculate [A] as a function of time, the results are as shown in Figure 2.7.

It is instructive to consider the initial rate of the reaction A $\longrightarrow$ B when the initial concentration of A varies. Figure 2.8 shows the variation in [A] for the cases where $[A]_o = 1.00$ M and $[A]_o = 2.00$ M when $k_1 = 0.050$ min^{-1} and k_{-1}

Table 2.2 Data for the reversible reaction A $\rightleftharpoons$ B.

Time (min)	[A], M
0	1.00
5	0.784
10	0.624
15	0.505
20	0.418
30	0.304
40	0.242
50	0.208
60	0.189
70	0.179
80	0.174

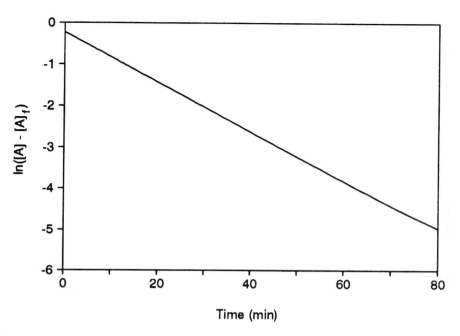

Figure 2.6   A rate plot for a reversible reaction using the conditions described in the text.

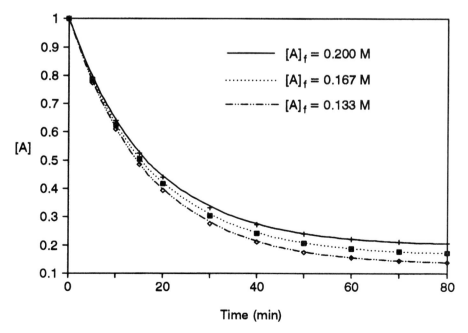

Figure 2.7   Variation in [A] when $[A]_f$ is in error by ± 0.033 **M** for the example described in the text.

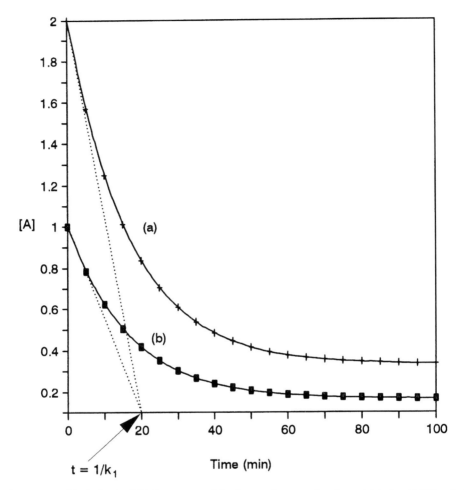

Figure 2.8 Variation in [A] for a reversible first-order reaction for which $k_1 = 0.050$ min^{-1} and $k_{-1} = 0.010$ min^{-1}. Curve (a) corresponds to $[A]_o = 2.00$ **M** and curve (b) corresponds to $[A]_o = 1.00$ **M**. Note that the initial rates extrapolate to a value of $1/k_1$.

$= 0.010$ min^{-1} and $K = 5.00$. When $[A]_o = 1.00$ **M**, $[A]_f = 0.167$ **M**, but when $[A]_o = 2.00$ **M**, $[A]_f = 0.333$ **M**. However, when the tangents are drawn for the two rate plots to indicate the initial rates, it is seen that they both have abscissa values of $1/k_1$. The initial rate of the reaction can be represented as shown in Eq. (2.68),

$$-\frac{d[A]}{dt} = k_1[A]_o - k_{-1}[B]_o$$

However, no B is initially present so we can write

$$-\int_{[A]_o}^{[A]} d[A] = k_1[A]_o \int_0^t dt$$

which yields on integration

$$[A]_o - [A] = k_1[A]_o t$$

At the beginning of the reaction, $[A] = 0$ so

$$[A]_o = k_1[A]_o t$$

or

$$t = 1/k_1 \tag{2.76}$$

It is apparent that the extrapolation of the *initial* rate to the time when $[A] = 0$ yields a time which is $1/k_1$, and this value is independent of the initial concentration of A.

The problem of reactions which do not go to completion is a frequently occurring one. We have shown here only the mechanics of dealing with a reversible system in which the reaction in each direction is first-order. Other cases that might arise are reversible second-order reactions, series reactions in which only one step is reversible, etc. These cases are quite complicated and their treatment is beyond the scope of this survey. However, many such systems have been elegantly described (Schmid and Sapunov, 1982). The interested reader is directed to these worked-out exercises in applied mathematics for more details.

2.6   AUTOCATALYSIS

For certain reactions, it is observed that the rate of the reaction increases as the reaction progresses. Such a situation occurs when a product acts as a catalyst for the reaction. Suppose the reaction

$$A \longrightarrow B$$

is first-order in A but that the reaction is catalyzed by B. As B is formed, the rate of the reaction will increase and the first-order plot will deviate from linearity with the slope of the line increasing, as shown in Figure 2.9.

Mathematically, this system can be described as

$$-\frac{d[A]}{dt} = k[A][B] \tag{2.77}$$

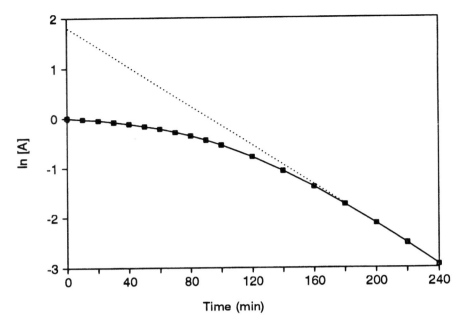

Figure 2.9 A plot of ln[A] vs. time for the autocatalytic process A ⟶ B. Conditions are $[A]_o = 1.00$ **M**, $[B]_o = 0.100$ **M**, and $k = 0.020$ min^{-1}.

Because of the stoichiometry of the reaction shown, the concentration of A reacted is equal to the concentration of B produced so that

$$[A]_o - [A] = [B] - [B]_o$$

Therefore, the concentration of B can be expressed as

$$[B] = [A]_o + [B]_o - [A] \tag{2.78}$$

which substituted into Eq. (2.77) gives

$$-\frac{d[A]}{dt} = k[A]([A]_o + [B]_o - [A]) \tag{2.79}$$

This equation can be written as

$$-\frac{d[A]}{[A]([A]_o + [B]_o - [A])} = k\,dt$$

It is apparent that this equation is very similar to that for the mixed second-order case shown in Eq. (2.6). We can write

$$\frac{1}{[A]([A]_o + [B]_o - [A])} = \frac{C_1}{[A]} + \frac{C_2}{[A]_o + [B]_o - [A]} \tag{2.80}$$

and

$$\frac{C_1}{[A]} + \frac{C_2}{[A]_o + [B]_o - [A]} = \frac{C_1([A]_o + [B]_o - [A]) + C_2[A]}{[A]([A]_o + [B]_o - [A])}$$

so that

$$C_1([A]_o + [B]_o - [A]) + C_2[A] = 1 \tag{2.81}$$

Expanding this expression yields

$$C_1[A]_o + C_1[B]_o - C_1[A] + C_2[A] = 1$$

When A has reacted completely,

$$C_1[A]_o + C_1[B]_o = 1$$

and

$$C_2[A] - C_1[A] = 0$$

Because [A] is not zero except after infinite time, $C_1 = C_2$ and

$$C_1([A]_o + [B]_o) = 1 \tag{2.82}$$

Therefore,

$$C_1 = \frac{1}{([A]_o + [B]_o)} = C_2$$

Therefore,

$$-\left[\frac{d[A]}{[A]([A]_o + [B]_o)} + \frac{d[A]}{([A]_o + [B]_o)([A]_o + [B]_o - [A])} \right] = k \, dt$$

This equation can be integrated to yield

$$\left[\frac{1}{[A]_o + [B]_o} \ln \frac{[A]_o}{[A]} + \frac{1}{[A]_o + [B]_o} \ln \frac{[A]_o + [B]_o - [A]}{[B]_o} \right] = kt \tag{2.83}$$

Upon substituting $[B] = [A]_o + [B]_o - [A]$ and simplifying, one obtains

$$\frac{1}{[A]_o + [B]_o} \ln \frac{[A]_o[B]}{[B]_o[A]} = kt \tag{2.84}$$

This equation can be rearranged to give

$$\ln \frac{[A]_o[B]}{[B]_o[A]} = ([A]_o + [B]_o)kt$$

or

$$\frac{[A]_o[B]}{[B]_o[A]} = \exp(([A]_o + [B]_o)kt) \tag{2.85}$$

Because

$$[A] = [A]_o + [B]_o - [B]$$

we can write

$$\frac{[A]_o[B]}{[A]_o + [B]_o - [B]} = [B]_o \exp(([A]_o + [B]_o)kt) \tag{2.86}$$

or

$$[A]_o[B] = [A]_o[B]_o \exp(([A]_o + [B]_o)kt) + [B]_o^2 \exp(([A]_o + [B]_o)kt) - [B][B]_o \exp(([A]_o + [B]_o)kt)$$

Rearranging and solving for $[B]$ gives

$$[B] = \frac{[A]_o[B]_o \exp(([A]_o + [B]_o)kt) + [B]_o^2 \exp(([A]_o + [B]_o)kt)}{[B]_o \exp(([A]_o + [B]_o)kt) + [A]_o} \tag{2.87}$$

Dividing each term in the numerator and denominator by the quantity $[B]_o \exp(([A]_o + [B]_o)kt)$ gives the final expression for $[B]$,

$$[B] = \frac{[A]_o + [B]_o}{1 + \dfrac{[A]_o}{[B]_o} \exp(-([A]_o + [B]_o)kt)} \tag{2.88}$$

From Eq. (2.85), we find that substitution for $[B]$ gives

$$\frac{[A]_o[B]}{[B]_o[A]} = \exp(([A]_o + [B]_o)kt) = \frac{[A]_o([A]_o + [B]_o - [A])}{[B]_o[A]}$$

and

$$\frac{[A]_o[B]}{[B]_o[A]} = \frac{[A]_o{}^2 + [A]_o[B]_o - [A]_o[A]}{[B]_o[A]}$$
(2.89)

Therefore,

$$[B]_o[A]\exp(([A]_o + [B]_o)kt) = [A]_o{}^2 + [A]_o[B]_o - [A]_o[A]$$

which can be written as

$$[B]_o[A]\exp(([A]_o + [B]_o)kt) - [A]_o{}^2 - [A]_o[B]_o + [A]_o[A] = 0$$
(2.90)

Solving this equation for [A] gives

$$[A] = \frac{[A]_o{}^2 + [B]_o[A]_o}{[A]_o + [B]_o \exp(([A]_o + [B]_o)kt)}$$
(2.91)

Dividing each term in the numerator and denominator by $[A]_o$ yields the equation in the form most often encountered.

$$[A] = \frac{[A]_o + [B]_o}{1 + \dfrac{[B]_o}{[A]_o} \exp(([A]_o + [B]_o)kt)}$$
(2.92)

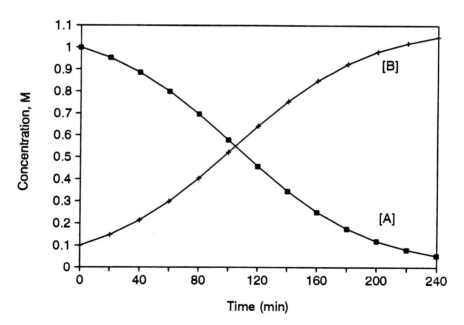

Figure 2.10   Concentration of A and B for the autocatalytic process, A ⟶ B. Conditions are $[A]_o = 1.00$ **M**, $[B]_o = 0.100$ **M**, and $k = 0.020$ min^{-1}.

Figure 2.10 shows the variation of [A] and [B] for the case where $[A]_o$ = 1.00 **M**, $[B]_o$ = 0.100 **M**, and k = 0.020 min^{-1}. The sigmoidal nature of the curves, which is characteristic of autocatalytic processes, is clearly visible. The nature of the curves suggests that the rate goes through a maximum at some [A], then decreases as the concentration gets lower.

It is of interest to note that a plot of ln [A] versus t shows first-order behavior at longer times even though the overall plot is curved. Figure 2.9 shows this behavior for the same initial data used to prepare Figure 2.10. From such a plot, it is possible to evaluate k, as is also shown in Figure 2.9 where the slope is $-k([A]_o + [B]_o)$.

2.7 EFFECT OF TEMPERATURE

In Chapter 1, the effect of temperature on reaction rate was illustrated by means of the Arrhenius equation. In most cases, a reaction can be studied conveniently over a rather narrow range of temperature, perhaps 30 to 40°C. However, for a very wide range of T, a plot of ln k versus T would not be linear.

If we consider the Arrhenius equation,

$$k = Ae^{-E/RT} \tag{2.93}$$

we can see why there is not a linear relationship between ln k and T over a great range of temperature. For example, if the activation energy is 100 kJ mol^{-1} and the temperature is 1000 K, k would have a value of about 6×10^{-6}A. At 2000 K, k is 2.45 $\times 10^{-3}$A, at 10,000 K, k = 0.300A, etc. It can be shown that at sufficiently high T, $k \longrightarrow$ A because RT $\longrightarrow \infty$ and $e^{-1/RT} \longrightarrow$ 1. In fact, a plot of k versus T is sigmoidal and k approaches A as an upper limit. In the usual temperature range studied, k increases with temperature in the way described earlier.

It is often stated as a general rule that the rate of a reaction doubles for a 10° rise in temperature. This can be examined easily by writing the Arrhenius equation for two temperatures as

$$E = \frac{RT_1T_2}{T_2 - T_1} \ln \frac{k_2}{k_1} \tag{2.94}$$

By choosing T_1 and T_2 so that they represent a 10° interval, we can evaluate ln (k_2/k_1) and hence k_2/k_1, as related to E. For example, if we take T_2 = 305 K and T_1 = 295 K and calculate k_2/k_1 for various values of E, we obtain the results shown in Figure 2.11. It is obvious that k_2/k_1 = 2 (the rate doubles) only if the activation energy is about 50 kJ mol^{-1}. On the other hand, if E is about 150 kJ mol^{-1}, k_2/k_1 = 7.4 if the temperature is increased from 295 to 305 K.

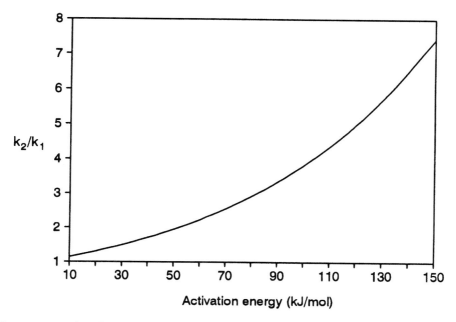

Figure 2.11   The effect of a temperature increase from 295 to 305 K on the ratio of k_2/k_1.

This behavior suggests that it would be of interest to examine the relationship between E, k_2/k_1, and the temperature interval because both E and the temperature where the 10° range occurs affect k_2/k_1. Eq. (2.94) can be written as

$$\frac{E(T_2 - T_1)}{R} = T_1 T_2 \ln \frac{k_2}{k_1} \tag{2.95}$$

For a specified value of E and T_1 and T_2 chosen so that $T_2 - T_1$ is 10 K, the left-hand side of Eq. (2.95) is constant, and hyperbolas result when $\ln (k_2/k_1)$ is plotted against $T_1 T_2$. In Figure 2.12, curves are shown that are obtained when k_2/k_1 is plotted versus the average temperature in a 10° interval for activation energies of 50, 75, and 100 kJ mol^{-1}. It is clear from this figure that k_2/k_1 has a value of 2 at different temperatures depending on the activation energy. At 300 K (interval 295 to 305 K), k_2/k_1 is approximately 2 when the activation energy is 50 kJ mol^{-1}, but it is approximately 3 if the activation energy is 75 kJ mol^{-1}, and it is approximately 4 if the activation energy is 100 kJ mol^{-1}. At intervals involving lower average temperatures, the effect is much greater, while for intervals involving higher temperatures, the effect is much less.

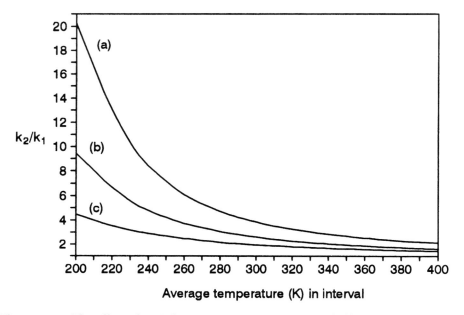

Figure 2.12   The effect of a 10 degree rise in temperature on k_2/k_1 for different activation energies. Curves (a), (b), and (c) correspond to activation energies of 100, 75, and 50 kJ/mol, respectively.

The effect of temperature on reaction rate was first observed over 100 years ago. Hood first noted that

$$\log k = \frac{-A}{T} + B \tag{2.96}$$

which we now write as

$$\ln k = \ln A - E_a / RT \tag{2.97}$$

For chemical equilibrium,

$$\frac{d \ln K}{dT} = \frac{\Delta E}{RT^2} \tag{2.98}$$

For the reaction

$$A + B \underset{k_{-1}}{\overset{k_1}{\rightleftharpoons}} C + D \tag{2.99}$$

the forward and reverse reactions have equal rates at equilibrium, so we can write

$$k_1[A][B] = k_{-1}[C][D]$$

Therefore, the equilibrium constant can be written as

$$K = \frac{k_1}{k_{-1}} = \frac{[C][D]}{[A][B]} \tag{2.100}$$

Substituting this result into Eq. (2.98) gives

$$\frac{d \ln k_1}{dT} - \frac{d \ln k_{-1}}{dT} = \frac{\Delta E}{RT^2}$$

From this equation, we can write

$$\frac{d \ln k_1}{dT} = \frac{E_1}{RT^2} \tag{2.101}$$

and

$$\frac{d \ln k_{-1}}{dT} = \frac{E_{-1}}{RT^2} \tag{2.102}$$

from which

$$E_1 - E_{-1} = \Delta E \tag{2.103}$$

with the energies involved being illustrated in Figure 2.13. As was shown in Chapter 1, a plot of $\ln k$ versus $1/T$ is linear with a slope of $-E/R$. However, for reactions which are studied over a very large range of temperature, the plots are not exactly linear as described earlier.

For some processes, the frequency factor is also a function of temperature so that

$$k = AT^n \, e^{-E/RT} \tag{2.104}$$

where n is usually an integer or half-integer. Therefore, a more accurate equation which is seldom used is

$$\ln k = \ln A + n \ln T - E/RT \tag{2.105}$$

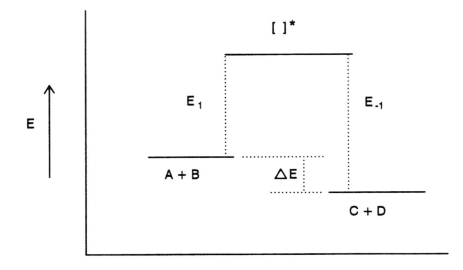

Reaction coordinate

Figure 2.13 Energy relationship for the reaction A + B ⟶ C + D. The activated complex or transition state is denoted as [ ]*.

The interpretation of E_1, E_{-1}, and ΔE is shown graphically in Figure 2.13. In this case, the activation energy for the forward reaction is E_1 while that for the reverse reaction is E_{-1}. From thermodynamics, we know that

$$\ln K = \frac{-\Delta G}{RT} = \frac{-\Delta H}{RT} + \frac{\Delta S}{R} \tag{2.106}$$

Writing a similar equation of this form for both the forward and reverse reaction and combining them with Eq. (2.98) enables us to show that*

$$\Delta H = E_1 - E_{-1} \tag{2.107}$$

Since the equilibrium constant for a reaction is related to ΔG,

$$\Delta G = -RT \ln K \tag{2.108}$$

or

$$K = e^{-\Delta G / RT} \tag{2.109}$$

*For details, see S. W. Benson, *The Foundations of Chemical Kinetics*, McGraw-Hill, New York, 1960, pp. 70–72. Note that Benson uses a mixture of subscripts in his Eq. (IV.3A.5), which makes it unclear what energies are involved.

By analogy, we can write

$$k_1 = A_1 \exp(-\Delta G_1^{\ddagger} / RT) \tag{2.110}$$

and

$$k_{-1} = A_{-1} \exp(-\Delta G_{-1}^{\ddagger} / RT) \tag{2.111}$$

By the principle of microscopic reversibility, we assume that the activated complex is the same regardless of which direction the reaction takes place. In such a case, it is reasonable to assume that $A_1 = A_{-1}$, and that the difference in rates of the forward and reverse reactions is due only to the difference in $\Delta G^{\ddagger}$ values. Therefore,

$$\frac{k_1}{k_{-1}} = \exp((\Delta S_1^{\ddagger} - \Delta S_{-1}^{\ddagger}) / R) \exp((-\Delta H_1^{\ddagger} + \Delta H_{-1}^{\ddagger}) / RT) \tag{2.112}$$

Kinetic studies generally deal with the forward direction, so

$$k_1 = A_1 \exp(\Delta S_1^{\ddagger} / R) \exp(-\Delta H_1^{\ddagger} / RT) \tag{2.113}$$

which can be written in logarithmic form as

$$\ln k_1 = \ln A_1 + \frac{\Delta S_1^{\ddagger}}{R} - \frac{\Delta H_1^{\ddagger}}{RT} \tag{2.114}$$

which is known as the Eyring equation. When $\ln k$ is plotted versus $1/T$, a line is obtained having a slope of $-\Delta H^{\ddagger}/R$. Once $\Delta H^{\ddagger}$ is known, $\Delta S^{\ddagger}$ can be calculated by means of Eq. (2.114). The entropy of activation is a useful property that is based on the choice of standard states. For a gas phase reaction in which a molecule $X - Y$ dissociates, $\Delta S^{\ddagger}$ would be expected to be positive. However, if the reaction takes place in solution and if the solvent is polar, dissociation of $X - Y$ into X^+ and Y^- followed by solvation of the ions could result in $\Delta S^{\ddagger}$ being negative. It should be noted that Eq. (2.112) applies strictly only to first-order processes. For example, see R. Schmid and V. N. Sapunov, *Non-formal Kinetics*, Verlag Chemie, Weinheim, 1982, p. 110. We will have more to say about the effects of temperature and solvation in later chapters.

REFERENCES FOR FURTHER READING

Benson, S. W. (1960) *The Foundations of Chemical Kinetics*, McGraw-Hill, New York, Chapter 3. Details of many reaction systems.

Frost, A. A., Pearson, R. G. (1961) *Kinetics and Mechanism*, 2d ed., Wiley, New York, Chapters 2, 3, and 8. Several methods of analyzing kinetic data are described.

Laidler, K. J. (1965) *Chemical Kinetics*, McGraw-Hill, New York, Chapter 1. A standard kinetics text dealing with gas phase reactions and reactions in solution.
Schmid, R., Sapunov, V. N. (1982) *Non-formal Kinetics*, Verlag Chemie,
Weinheim. A marvelous book showing the applied mathematics associated with many complex cases.

PROBLEMS

1. The reaction of NO(g) with $Br_2(g)$ is believed to take place in two steps:

$$NO(g) + Br_2(g) \underset{k_{-1}}{\overset{k_1}{\rightleftharpoons}} ONBr_2(g) \qquad \text{(fast)}$$

$$ONBr_2(g) + NO(g) \overset{k_2}{\longrightarrow} 2\ ONBr(g) \qquad \text{(slow)}$$

The $ONBr_2(g)$ is unstable. On the basis of this information, write the rate law
expected for the reaction. Obtain the rate law describing the concentration of
$ONBr_2(g)$ with time and derive the final rate law for the reaction.

2. The reaction A $\longrightarrow$ P produces the data shown below.

Time (min)	0	20	40	60	80	100	120
[A] (**M**)	0.800	0.709	0.557	0.366	0.201	0.096	0.042

Plot [A] versus time and tell as much as you can about the mechanism of the
reaction from that graph. Now determine the rate law for the reaction and
evaluate the rate constant(s).

3. For the reacting system

$$X \underset{k_{-1}}{\overset{k_1}{\rightleftharpoons}} Y,$$

the following data were obtained when no Y was present initially and two
starting concentrations of X were used.

Time (hr)	0	10	20	30	40	50	60
[X] (**M**)	0.600	0.374	0.250	0.182	0.145	0.125	0.113
[X] (**M**)	1.200	0.750	0.501	0.365	0.290	0.250	0.227

Write the rate equation for the change in [X] with time. Make appropriate substitutions and determine the final rate equation and then integrate it. Use the data above to determine k_1.

4. Suppose a dimer, A_2, reacts by first dissociating into monomers, then it is transformed into B.

$$A_2 \underset{k_{-1}}{\overset{k_1}{\rightleftharpoons}} 2\ A \xrightarrow{k_2} B$$

Assume that a steady-state concentration of A is maintained and derive the expression for the rate of disappearance of A_2. Integrate this expression to obtain the integrated rate law.

5. ^{64}Cu undergoes radioactive decay by β^+ emission to produce ^{64}Zn and β^- emission and by electron capture simultaneously to produce ^{64}Ni. If the half-life of ^{64}Cu is 12.8 years, obtain an expression for the amounts of ^{64}Zn and ^{64}Ni at any time, t.

6. ^{38}S decays by β^- emission to ^{38}Cl with a half-life 2.87 hrs. The ^{38}Cl decays by β^- emission to ^{38}Ar with a half-life of 37.3 min. Obtain an expression for the amount of each nuclide as a function of time.

7. In reference to Problem 6, determine the maximum number of ^{38}Cl atoms that is ever present if the original sample of ^{38}S contains 10^4 atoms.

8. Consider the elementary steps in the reaction of NO and F_2.

$$NO + F_2 \xrightarrow{k_1} ONF + F$$

$$NO + F \xrightarrow{k_2} ONF$$

Derive the rate laws that result from the conditions
(a) $k_1 \gg k_2$, (b) $k_2 \gg k_1$, and (c) with $k_1 \approx k_2$.

9. Cadmium-117 undergoes β^- decay with a half-life of 2.4 hrs to ^{117}In, which undergoes β^- decay with a half-life of 42 min to ^{117}Sn, which is stable. (a) If a sample of initially pure ^{117}Cd contains 1.50×10^6 atoms, how many atoms will remain after 4.00 hrs? (b) How many atoms of ^{117}In will be present after 4.00 hours? (c) How many atoms of ^{117}Sn will be present after this time?

chapter 3

TECHNIQUES AND METHODS

In Chapter 2, several kinetic schemes were examined in detail. While the mathematical apparatus was developed for several cases, little was said about other methods used in kinetic studies or about experimental techniques. In this chapter, we will describe some of the methods of kinetics that do not make use of the integrated rate laws. Some of the experimental techniques that do not make use of the classical determination of concentration as a function of time will also be described. Bernasconi, Editor (1986) *Investigation of Rates and Mechanisms of Reactions* contains two parts of the series *Techniques of Chemistry*, (Weissberger, Series editor) which describes all phases of kinetics techniques. Part I, *General Considerations and Reactions at Conventional Rates*, would be especially valuable for a study of kinetic methods. Part II, *Investigation of Elementary Reaction Steps in Solution and Fast Reaction Techniques*, deals with additional aspects of solution kinetics. These books should be consulted for full discussions of kinetics methods.

3.1 CALCULATING RATE CONSTANTS

One of the standard ways of examining data from a kinetic analysis is that of preparing a table of the results. In this method, the data analyzed by a particular rate law are used to calculate the rate constant for each (concentration, time) data pair. In Section 2.1, a kinetic study of the hydrolysis of ethyl acetate with sodium hydroxide was described. In that experiment, the initial concentration of NaOH was 0.0100 **M** (a) and the initial concentration of $CH_3COOC_2H_5$ was 0.00580 **M** (b). Therefore, $(a - b) = 0.00420$ **M** and $1/(a - b) = 238.1$. Using these values and the kinetic data shown in Table 2.1, the results shown in Table 3.1 were obtained.

Several factors should be noted from the results shown in Table 3.1. First, the calculated k values for times of 1 and 3 minutes deviate rather significantly from the values at longer times. This is undoubtedly due to the difficulties in mixing solutions and obtaining homogeneous aliquots at very short times after mixing the solutions and the fact that the sampling time itself is a significant fraction of

Table 3.1   Calculated rate constants for the hydrolysis of $CH_3COOC_2H_5$ using the experimental data shown in Table 2.1.

Time (min)	$\dfrac{(a-x)}{(b-x)}$	$\dfrac{b(a-x)}{a(b-x)}$	$\ln\dfrac{b(a-x)}{a(b-x)}$	$\dfrac{1}{(a-b)}\ln\dfrac{b(a-x)}{a(b-x)}$	k (M^{-1} min^{-1})
1	1.83	1.06	0.0596	14.2	14.2
3	1.90	1.10	0.0971	23.1	7.71
5	2.10	1.22	0.197	46.9	9.38
10	2.50	1.45	0.372	88.6	8.86
20	3.38	1.96	0.673	160	8.01
35	5.57	3.24	1.17	279	7.98
55	12.8	7.37	2.00	475	8.65
75	25.3	14.7	2.68	639	8.52

the measured time. The data used were obtained from an actual experimental kinetics run. Second, there is no trend in the calculated k values, which would suggest that an incorrect rate law has been used to correlate the data.

In Section 1.3, we described the difficulties in analyzing data where errors in the data make it difficult to determine the applicable concentration function. Using the method of calculated rate constants may make it impossible to distinguish between experimental errors in the data while graphical presentation of the data *may* reveal a trend or curvature, which *suggests* that another rate law is applicable. Finally, when the calculated rate constants are displayed as in Table 3.1, it is usually impossible to detect a trend in the values unless the reaction has been studied over a large fraction of reaction. In all cases where it is possible to do so, a reaction should be studied over several half-lives in order to obtain data which are amenable to kinetic analysis. The data shown in Table 3.1 indicate that studying the reaction for only a few minutes would not have made it possible to say much about the kinetics of the reaction.

3.2   METHOD OF HALF-LIVES

In Section 2.2, the equation

$$t_{1/2} = \frac{2^{n-1} - 1}{(n-1)\,k[A]_0^{\,n-1}} \tag{3.1}$$

was derived. This equation can be written as

$$t_{1/2} = \frac{1}{[A]_0^{\,n-1}} \cdot \frac{2^{n-1} - 1}{(n-1)k} = \frac{1}{[A]_0^{\,n-1}}\, f(k,n) \tag{3.2}$$

Taking the logarithm of both sides of Eq. (3.2) gives

$$\ln t_{1/2} = -(n-1) \ln[A]_\circ + \ln f(k,n) \tag{3.3}$$

Now, because k and n are constants, the last term is constant if the experimental conditions are not changed. Therefore, if several reactions are carried out at $[A]_\circ$, $[A]_\circ/2$, $[A]_\circ/4$, etc., and the half-life for the reaction is determined in each case, a plot of $\ln [A]_\circ$ versus $\ln t_{1/2}$ will yield a straight line of slope $-1(n-1)$.

   If the reaction is carried out using two different $[A]_\circ$ values, Eq. (3.3) gives

$$\frac{(t_{1/2})_1}{(t_{1/2})_2} = \frac{\dfrac{1}{([A]_\circ^{n-1})_1}}{\dfrac{1}{([A]_\circ^{n-1})_2}} = \frac{([A]_\circ^{n-1})_2}{([A]_\circ^{n-1})_1} \tag{3.4}$$

Taking logarithms of both sides of the equation, we obtain

$$\ln \frac{(t_{1/2})_1}{(t_{1/2})_2} = \ln \frac{([A]_\circ^{n-1})_2}{([A]_\circ^{n-1})_1} \tag{3.5}$$

which can be written as

$$\ln(t_{1/2})_1 - \ln(t_{1/2})_2 = (n-1) \ln([A]_\circ)_2 - (n-1) \ln([A]_\circ)_1$$

or

$$\ln(t_{1/2})_1 - \ln(t_{1/2})_2 = (n-1)\{\ln([A]_\circ)_2 - \ln([A]_\circ)_1\}$$

This equation can be rearranged to give

$$\frac{\ln(t_{1/2})_1 - \ln(t_{1/2})_2}{\ln([A]_\circ)_2 - \ln([A]_\circ)_1} = n - 1$$

which gives n as

$$n = \frac{\ln(t_{1/2})_1 - \ln(t_{1/2})_2}{\ln([A]_\circ)_2 - \ln([A]_\circ)_1} + 1 \tag{3.6}$$

This equation shows that determining the half-life at two initial concentrations of A permits calculation of n. While this quick, approximate method is valid, it is not generally as accurate as more detailed methods of data analysis.

3.3   INITIAL RATES

For a reaction involving a single reactant, the rate, R, can be written as

$$R = k[A]^n \qquad (3.7)$$

Therefore, taking logarithms of both sides of the equation gives

$$\ln R = \ln k + n \ln[A] \qquad (3.8)$$

For a series of initial concentrations, the concentration of A varies with time as shown in Figure 3.1. The initial rates are determined from the slopes of the tangents. From the initial rates determined from the slopes, the data shown in Table 3.2 are derived.

Now, a plot of $\ln R$ versus $\ln[A]_o$ shown in Figure 3.2 gives a linear relationship with a slope of n, the order of the reaction with respect to A. For this illustration, calculated data for a reaction having $n = 1$ and a rate constant of 0.020 min^{-1} were used. Linear regression of the data above, which were determined graphically from Figure 3.1, yield a slope of 0.97, which is the reaction order, and an intercept of -9.94. Since the intercept is $\ln k$, this value corresponds to $k = 0.0195$ min^{-1}, which is almost exactly the same as the value of

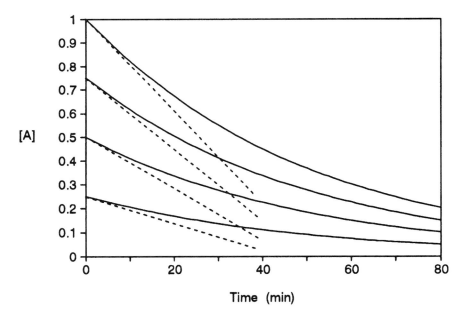

Figure 3.1   The method of initial rates.

Table 3.2   Data derived from the method of initial rates based
on the rate plots shown in Figure 3.1.

$[A]_o$ **M**	ln $[A]_o$	Initial rate **M^{-1} min^{-1}**	ln (rate)
1.00	0.000	0.0190	−3.96
0.75	−0.288	0.0150	−4.20
0.50	−0.693	0.0101	−4.60
0.25	−1.39	0.0050	−5.30

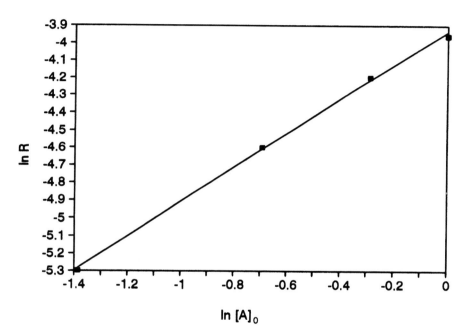

Figure 3.2   Plot of the logarithm of initial rate versus the logarithm of the initial concentration.

0.020 min^{-1} used in the calculations. The slight difference is due to the fact that
the slopes giving the initial rates were determined graphically by drawing the
tangents to the concentration versus time curves.

Initial rates can be used in another way. For example, suppose that a chemical reaction,

$$aA + bB \longrightarrow \text{Products} \tag{3.9}$$

follows a rate law which can be written in terms of concentrations as

$$\text{Rate} = -\frac{d[A]}{dt} = k[A]^n[B]^m \tag{3.10}$$

If the reaction is carried out using known initial concentrations of A and B, the initial rate can be determined graphically as shown in Figure 3.1. This procedure determines the initial rate, $(-d[A]/dt)_i$, which is given by

$$\left(-\frac{d[A]}{dt}\right)_{i1} = k[A]_o^n[B]_o^m \tag{3.11}$$

Now the process can be repeated using different concentrations of A and B, which we will take to be $2[A]_o$ and $2[B]_o$, and the initial rate determined is

$$\left(-\frac{d[A]}{dt}\right)_{i2} = k(2[A]_o)^n(2[B]_o)^m \tag{3.12}$$

Since k is a constant, a ratio of the initial rates for the two sets of conditions gives

$$\frac{\left(-\dfrac{d[A]}{dt}\right)_{i2}}{\left(-\dfrac{d[A]}{dt}\right)_{i1}} = \frac{k(2[A]_o)^n(2[B]_o)^m}{k[A]_o^n[B]_o^m} = 2^n \times 2^m = 2^{n+m} \tag{3.13}$$

Therefore, the overall order, $n + m$, can easily be determined from the initial rates obtained at different starting concentrations of the reactants.

3.4   FLOODING

For a reaction such as that shown in Eq. (3.9), the rate law can be written as

$$-\frac{d[A]}{dt} = k[A]^n[B]^m \tag{3.14}$$

If the concentration of B is made large with respect to that of A, the concentration of B will not change significantly while the concentration of A changes by an amount x. Therefore,

$$-\frac{d[A]}{dt} = \frac{dx}{dt} = k([A]_o - x)^n[B]^m \tag{3.15}$$

Since [B] is essentially a constant, we can write

$$\frac{dx}{dt} = k'([A]_o - x)^n \tag{3.16}$$

This expression can be treated by the integral methods described in Chapter 2 to determine n. The procedure can be repeated making $[A]_o$ large compared to $[B]_o$ so that m can be determined.

Many reactions in aqueous solutions or those in which OH^- or H^+ is a reactant fit this situation. For example, in Chapter 1 the reaction of $(CH_3)_3CBr$ with OH^- in basic solution was described. Under these conditions, the concentration of OH^- is sufficiently large that the reaction appears to be first order in $(CH_3)_3CBr$, but it is actually a pseudo first-order process. Many hydrolysis reactions appear to be independent of $[H_2O]$ only because water is present in such a large excess.

3.5   LOGARITHMIC METHOD

Suppose that a reaction which can be shown as

$$aA + bB \longrightarrow Products \tag{3.17}$$

follows the rate law

$$R = k[A]^n[B]^m \tag{3.18}$$

If the reaction is run under two sets of conditions and the initial concentration of B is kept constant but the initial concentration of A is varied, the ratio of the rates will be

$$\frac{R_1}{R_2} = \frac{k[A]_1^n}{k[A]_2^n} \tag{3.19}$$

Taking logarithms (either natural or common) of both sides of the equation gives

$$\log \frac{R_1}{R_2} = n \log \frac{[A]_1}{[A]_2}$$

so that

$$n = \frac{\log(R_1 / R_2)}{\log([A]_1 / [A]_2)} \tag{3.20}$$

This procedure can be repeated with the initial concentration of A kept constant to determine m in a similar way by varying the starting concentration of B.

A somewhat better way is to run the reaction using several starting concentrations of A while keeping the concentration of B constant and to make a graph of the points obtained from the ratios R_i/R_{i+1} and $[A]_i/[A]_{i+1}$. Taking logarithms of the ratios gives

$$\log\left(\frac{R_i}{R_{i+1}}\right) = n \log\left(\frac{[A]_i}{[A]_{i+1}}\right) \tag{3.21}$$

The slope of the plot of $\log(R_i/R_{i+1})$ versus $\ln([A]_i/[A]_{i+1})$ will be equal to n, the order of the reaction with respect to A. The procedure can be repeated to find m, the order of the reaction with respect to B.

The reaction

$$S_2O_8^{2-} + 3\ I^- \longrightarrow I_3^- + 2\ SO_4^{2-} \tag{3.22}$$

is an interesting one because the rate law is not indicated by the balanced equation. It can be studied kinetically by monitoring the production of I_3^-, which gives the familiar blue color with starch. Because I_3^- oxidizes $S_2O_3^{2-}$ by the reaction

$$I_3^- + 2\ S_2O_3^{2-} \longrightarrow 3\ I^- + S_4O_6^{2-} \tag{3.23}$$

the amount of I_3^- produced can be determined by using a known concentration of $S_2O_3^{2-}$. When the $S_2O_3^{2-}$ is exhausted, the I_3^- produced interacts with the starch to produce a blue color. In this way, the concentration of $S_2O_8^{2-}$ reacted can be determined indirectly.

This reaction can be used to illustrate the application of the logarithmic method. In this study, the first run had an initial concentration of $S_2O_8^{2-}$ and I^- of 0.050 **M**. The initial rate of consumption of $S_2O_8^{2-}$ was found to be 4.4×10^{-5} **M** sec^{-1}. In the second run, the concentration of $S_2O_8^{2-}$ was 0.050 **M** while that of I^- was 0.100 **M**. In this case, the initial rate of $S_2O_8^{2-}$ consumption was 8.6×10^{-5} **M** sec^{-1}. In the final run, the concentration of $S_2O_8^{2-}$ was 0.100 **M** while the concentration of I^- was 0.050 **M** and the initial rate was 8.9×10^{-5} **M** sec^{-1}. Using Eq. (3.20), we obtain

$$n = \frac{\log(4.4 \times 10^{-5}/8.9 \times 10^{-5})}{\log(0.050/0.100)} = \frac{-0.306}{-0.301} = 1.0$$

and

$$m = \frac{\log(4.4 \times 10^{-5}/8.6 \times 10^{-5})}{\log(0.050/0.100)} = \frac{-0.291}{-0.301} = 0.97$$

Consequently, we can conclude that the reaction is first-order in $S_2O_8^{2-}$ and first-order in I^-, which reinforces the conclusion that the rate law must be deduced experimentally. Having determined the rate law, the rate constant can be found to be equal to 1.7×10^{-2} **M** sec^{-1}.

3.6   EFFECTS OF PRESSURE

The principle of Le Chatelier enables us to predict the effects of changing variables on a system at equilibrium. For example, increasing the temperature causes the system to shift in the endothermic direction. Likewise, increasing the pressure causes the system to shift in the direction corresponding to the smaller volume. In chemical reactions, we have repeatedly assumed that a small but essentially constant concentration of the activated complex is in equilibrium with the reactants. In Section 2.7, we dealt with the effect of temperature on the rate constant, but it should also be apparent that pressure can affect the value of k if the transition state occupies a different volume than that of the separate reactant species. As will be discussed in Chapter 5, the effect of the *internal* pressure caused by the solvent affects the rate of a reaction in much the same way as does the *external* pressure.

For a process that takes place in solution at constant temperature, we can write

$$\left(\frac{\partial G}{\partial P} \right)_T = V \tag{3.24}$$

where V is the partial molar volume. For a chemical reaction, the free energy of activation can be written as

$$\Delta G^{\ddagger} = G^{\ddagger} - \Sigma G_R \tag{3.25}$$

where $G^{\ddagger}$ is the free energy of the activated complex and ΣG_R represents the sum of the molar free energies of the reactants. Since the *volume of activation* is given by

$$\Delta V^{\ddagger} = V^{\ddagger} - \Sigma V_R \tag{3.26}$$

we can write

$$\left(\frac{\partial G^{\ddagger}}{\partial P} \right)_T = V^{\ddagger} - \Sigma V_R = \Delta V^{\ddagger} \tag{3.27}$$

Strictly speaking, the concentrations of reactants should be expressed in units which are independent of the changes in volume produced by pressure. Since volumes of liquids change due to compressibility, molality or mole fraction should be used to measure concentrations rather than molarity.

For a reaction carried out at constant temperature, we can use

$$\Delta G^{\ddagger} = -RT \ln K^{\ddagger} \tag{3.28}$$

where $K^{\ddagger}$ is the equilibrium constant for the formation of the transition state. The equation

$$\frac{\partial \ln k}{\partial T} = -\frac{E_a}{RT} \tag{3.29}$$

can be transformed into the equation

$$\left(\frac{\partial \ln k}{\partial P} \right)_T = -\frac{\Delta V^{\ddagger}}{RT} \tag{3.30}$$

Therefore, using derivatives and solving for $\Delta V^{\ddagger}$ gives

$$\Delta V^{\ddagger} = -RT \frac{d \ln k}{dP} \tag{3.31}$$

which can be written as

$$d \ln k = -\frac{\Delta V^{\ddagger}}{RT} dP$$

By integration, we obtain

$$\ln k = -\frac{\Delta V^{\ddagger}}{RT} \cdot P + \text{Constant} \tag{3.32}$$

This equation indicates that a plot of $\ln k$ versus P should be linear with a slope of $-\Delta V^{\ddagger}/RT$. While such plots are sometimes approximately linear, they often exhibit curvature, which indicates that $\Delta V^{\ddagger}$ is pressure dependent. To deal with this situation, we need either a theoretical approach to determine $\Delta V^{\ddagger}$ or a graphical procedure. The latter is the usual way to determine $\Delta V^{\ddagger}$ when $\ln k$ is written as

$$\ln k = a + bP + cP^2 \tag{3.33}$$

Then, from Eqs. (3.32) and (3.33),

$$-\frac{\Delta V^{\ddagger}}{RT} P + \text{Constant} = a + bP + cP^2 \tag{3.34}$$

and equating terms in P gives

$$-\frac{\Delta V^{\ddagger}}{RT}P = bP \qquad (3.35)$$

or

$$\Delta V^{\ddagger} = -bRT \qquad (3.36)$$

It is instructive at this point to consider the magnitude of the effect produced by pressure.

If we consider P – V work to be given by $P\Delta V$, a change in volume of 10 cm³/mol (0.010 l/mol) by a pressure of 1000 atm would produce

$$1000 \text{ atm} \times 0.010 \text{ l/mol} = 10 \text{ l atm/mol or } 1.01 \text{ kJ/mol}$$

(1 l atm/mol = 101 J/mol). This amount of energy would be about equal to that corresponding to a low temperature. Therefore, in order to accomplish a change that can be brought about by a modest change in temperature, an enormous change in pressure is required. Consequently, pressure effects are usually determined for reactions studied at several kbar (1 bar = 0.98692 atm). When pressure in the range up to 10 kbar is used, typical $\Delta V^{\ddagger}$ for reactions are usually in the range of ±25 cm³/mol.

The interpretations of volumes of activation are not always unambiguous, but generally if $\Delta V^{\ddagger}$ is negative, the rate of reaction increases as pressure is increased. This signifies that the transition state occupies a smaller volume than the reactants. As a general rule, the formation of a bond (associative mechanism) causes a $\Delta V^{\ddagger}$ change of perhaps –5 to –15 cm³/mol, while the breaking of a bond (dissociative mechanism) causes a $\Delta V^{\ddagger}$ change of +5 to +15 cm³/mol. However, a bond-breaking step in which ions are formed leads to a $\Delta V^{\ddagger}$ of –20 to –40 cm³/mol. The reason for this rather large negative value is that ions are strongly solvated, which leads to a compacting and ordering of the solvent surrounding the ions. Therefore, $\Delta V^{\ddagger}$ is made up of two parts: (1) an intrinsic part, which depends on the changes in molecular dimensions and (2) a solvation part, which depends on the extent of solvation.

$$\Delta V^{\ddagger} = \Delta V^{\ddagger}_{int} + \Delta V^{\ddagger}_{solv} \qquad (3.37)$$

If desolvation occurs as the transition state is formed, $\Delta V^{\ddagger}_{solv}$ will be positive. If forming the transition state involves forming ions, $\Delta V^{\ddagger}_{solv}$ will be negative because of the ordering of the solvent in the vicinity of the charged ions. This phenomenon is known as *electrostriction.*

Pressure studies are conducted in order to obtain mechanistic information in a variety of reactions. One reaction in which rate studies at high pressure have yielded considerable information is a linkage isomerization reaction which has been known for many years,

$$[Co(NH_3)_5ONO]Cl_2 \longrightarrow [Co(NH_3)_5NO_2]Cl_2 \qquad (3.38)$$

This reaction has been studied thermally and photochemically both in solution and in the solid state. The reaction takes place rapidly which is uncharacteristic of most *substitution* reactions of Co^{3+} complexes. A simplistic view of this process suggests that the Co—ONO bond could break and then the nitrite ion could reattach by bonding through the nitrogen atom. The other possibility is that the Co—ONO bond does not actually break but that it merely "slides" to form the Co—NO$_2$ linkage.

$$(3.39)$$

Mares, Palmer, and Kelm (1978) studied the rates of linkage isomerization in $[Co(NH_3)_5ONO]^{2+}$ and the analogous Rh^{3+} and Ir^{3+} complexes under high pressure in aqueous solution. It was found that the linkage isomerization takes place more rapidly at high pressures. From the pressure effects on the rate constants, the volume of activation was determined to be as follows for the complexes of different metal ions: Co^{3+}, -6.7 ± 0.4; Rh^{3+}, -7.4 ± 0.4; Ir^{3+}, -5.9 ± 0.6 cm^3 mol^{-1}. These values suggest that the transition state occupies a smaller volume than the reactants. This is inconsistent with a transition state $[M^{3+} NO_2^-]$ formed by breaking the M—ONO bond. It is generally accepted that this reaction does not take place by a bond-breaking–bond-making process. Furthermore, it is also known that the rate of linkage isomerization in $[Co(NH_3)_5ONO]^{2+}$ is independent of the concentration of NO_2^- in the solution, which indicates that the NO_2^- never leaves the coordination sphere. Support for the structure of the transition state involving the multiply bonded NO_2^- ion has been obtained by studying the reaction photochemically in the solid state and quenching the solid to very low temperature. Infrared spectra of the material shows new bands that are not characteristic of either Co—ONO or Co—NO$_2$ linkages but that were believed to be due to bonding in

Clearly, the study of linkage isomerization under high pressure has contributed to the understanding of the reaction mechanism.

3.7 FLOW TECHNIQUES

When reactants are mixed, there is some finite time necessary for them to form a steady state amount of the activated complex. This time is usually very short on the time scale used to study the reaction. In this period, often called a *transient* or *pre-steady state* period, the kinetic rate laws developed earlier do not apply and different experimental techniques must be used to study such processes. One technique, developed in 1923 by Hartridge and Roughton for the study of the reaction between hemoglobin and oxygen, makes use of a *continuous-flow* system. The two reacting solutions were forced under constant pressure into a mixing chamber as shown in Figure 3.3. After the liquids mix and the reaction starts, the mixture flows out of the mixing chamber to a point where a measuring device is located. A suitable measuring device for many reactions is a spectrophotometer to determine the concentration of a reactant or product from absorption measurements. The length of time that the reaction has been taking place is determined by the distance from the mixing chamber to the observation point. Calculation of the reaction time is possible by making use of the flow rate of the reactants and the diameter of the tube from the mixing chamber. Presently used continuous flow apparatus can study reactions fast enough to have a half-life of 1 ms. Many types of apparatus have been developed, and different methods of introducing the samples have been used.

In the *stopped-flow* technique, the solutions are forced from syringes into a mixing chamber. After a very short period of flow, perhaps a few ms, the flow is stopped suddenly when the observation cell is filled by an opposing piston, which is linked to a sensing switch that triggers the measuring device (Figure 3.4). Small volumes of solutions are used, and the kinetic equations are equivalent to those used in conventional methods where concentration and time are measured. Commercial stopped-flow apparatus is available with several modifications in

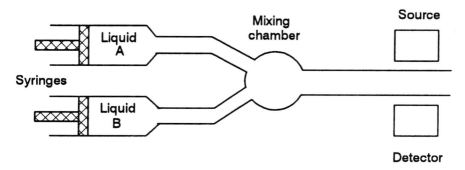

Figure 3.3 A continuous flow system with the source and detector of the spectrophotometer shown.

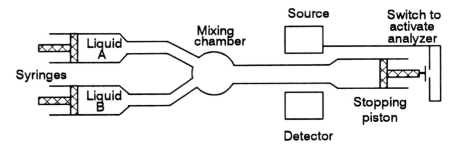

Figure 3.4   A stopped flow system with the source and detector of the spectrophotometer shown.

design. Both stopped-flow and continuous-flow techniques are capable of studying reactions with half-lives in the order of a few milliseconds.

3.8   TRACER METHODS

While kinetic methods can yield a great deal of useful information for interpreting reaction mechanisms, some questions may still remain. For example, a kinetic study on the hydrolysis of $CH_3COOC_2H_5$ was described in Chapter 2. The reaction was found to obey second-order mixed kinetics, but there is still a question to be answered: Which C—O bond breaks?

$$CH_3 - \overset{\overset{\displaystyle O}{\|}}{C} - O - C_2H_5 \qquad or \qquad CH_3 - \overset{\overset{\displaystyle O}{\|}}{C} - O - C_2H_5$$

In other words, does the oxygen atom in the ester linkage show up in the acid or in the alcohol? This question can be answered only if the oxygen atom is made distinguishable from those in the bulk solvent, water. The way to do this is to study the hydrolysis of $CH_3COOC_2H_5$ containing ^{18}O in that position. Then, hydrolysis of the ester will produce different products depending on which bond breaks.

$$CH_3 - \overset{\overset{\displaystyle O}{\|}}{C} - O^{18} - C_2H_5 \qquad or \qquad CH_3 - \overset{\overset{\displaystyle O}{\|}}{C} - O^{18} - C_2H_5$$
$$HO - H \qquad\qquad\qquad\qquad H - OH$$

Case I, ^{18}O in the alcohol   Case II, ^{18}O in the acid

In this reaction, the ^{18}O is found in the alcohol, showing that Case I is correct. An *insertion* reaction in a coordination compound can be written as

$$L_nM-X+Y \longrightarrow L_nM-Y-X \qquad (3.40)$$

where M is a metal atom or ion, n is the number of ligands of type L, and X and Y are other ligands. A reaction of this type for which tracer studies have yielded important mechanistic information is the CO insertion reaction of $[Mn(CO)_5CH_3]$.

$$\text{(structure)} + CO \longrightarrow \text{(structure)} \qquad (3.41)$$

For this insertion reaction, it might be assumed that it is the added CO that is being inserted in the Mn—CH_3 bond. However, when the CO being added is ^{14}CO, it is found that the ^{14}C is not found in the Mn—$COCH_3$ group. The reaction actually proceeds by a first step involving a group transfer reaction,

$$\text{(structure)} \longrightarrow \text{(structure)}$$

followed by coordination of the ^{14}CO.

$$\text{(structure)} + {}^{14}CO \longrightarrow \text{(structure)} \qquad (3.42)$$

In this reaction, the CO that is added to the complex is not the one that becomes inserted into the Mn—CH_3 bond.

The decomposition of NH_4NO_3 under carefully controlled conditions follows the equation

$$NH_4NO_3 \longrightarrow N_2O + 2H_2O \qquad (3.43)$$

Because the structure of N_2O is linear with a nitrogen atom in the middle,

$$\overline{N} = N = \overline{O}$$

the nitrogen atoms are not in equivalent positions. It is reasonable to wonder which nitrogen atom in N_2O came from NH_4^+ and which came from NO_3^-. In this case, using $^{15}NH_4NO_3$ and analyzing the N_2O produced shows that the product is ^{15}NNO. Therefore, the nitrate is not totally decomposed and one of the oxygen atoms is found attached to the same nitrogen atom that it was initially bonded to. The terminal nitrogen comes from the NH_4^+. A mechanism for the decomposition that is consistent with these observations can be shown as

$$NH_4^+NO_3^- \xrightarrow{-H_2O} [H_2N-NO_2] \xrightarrow{-H_2O} N=N=O \qquad (3.44)$$

A reaction that appears unusual when it is at first examined is

$$[Co(NH_3)_5OH]^{2+} + N_2O_3 \longrightarrow [Co(NH_3)_5ONO]^{2+} + HONO \qquad (3.45)$$

This reaction takes place quickly, which is itself part of the problem because substitution reactions of Co^{3+} low-spin complexes are very slow. The crystal field stabilization energy in such complexes is 24Dq, and substitution reactions usually take place by a dissociative pathway. Further, the isomer produced, $[Co(NH_3)_5ONO]^{2+}$, is the *less* stable isomer, with $[Co(NH_3)_5NO_2]^{2+}$ being the more stable product (see Section 3.6). It seems unlikely that the Co—OH bond would be so easily broken in this case because it is kinetically rather inert in others. These observations led to a tracer study of this reaction in which $[Co(NH_3)_5{}^{18}OH]^{2+}$ was used. In that case, the ^{18}O was found in the product, $[Co(NH_3)_5{}^{18}ONO]^{2+}$, indicating that the Co—O bond is not broken and the reaction takes place by a mechanism like

$$(NH_3)_5Co-{}^{18}O-H \longrightarrow [Co(NH_3)_5{}^{18}ONO]^{2+} + HONO$$
$$ON-NO_2$$

This reaction is actually the reaction of a coordinated ligand rather than a substitution reaction. A similar result has been found in the acid hydrolysis of the carbonato complex, $[Co(NH_3)_5CO_3]^+$.

A classic example of the use of isotopically labelled compounds in organic chemistry is the demonstration of the benzyne intermediate by J. D. Roberts *et al.* (1956). The reaction of chlorobenzene with amide ion produces aniline. This reaction was studied using ^{14}C at the 1-position.

$$\sim 50\% \qquad \sim 50\% \qquad \qquad \textbf{(3.46)}$$

It is believed that the reaction takes place by the formation of the benzyne intermediate,

$$\textbf{(3.47)}$$

Attack by NH_3 on the benzyne intermediate is about equally probable in forming a C—N bond at either end of the triple bond. Therefore, the product distribution is approximately 50% of either ^{14}C—N or ^{12}C—N bonds. Labelling the 1-^{14}C-chlorobenzene produces results which give a way to indicate to which carbon atom the NH_2 group attaches, and it makes possible a knowledge of the type of intermediate formed. While only a few examples have been cited here, the use of tracers in elucidating reaction mechanisms has been of great value. In many cases, the results obtained are simply not obtainable by other techniques.

3.9   KINETIC ISOTOPE EFFECTS

Molecules that are chemically identical but that contain different isotopes react at different rates. It is the difference in rates of electrolysis that allows D_2O to be obtained by the electrolysis of water even though the relative abundance of D to H is 1:6000. This is known as the *kinetic isotope effect*. A *primary* kinetic isotope effect occurs when isotopic substitution has taken place so that the bond being broken directly involves different isotopes. A mathematical treatment of isotope effects is beyond the scope of this book, but we can see how they arise in a straightforward way.

It is known that the greater the relative difference in mass, the greater the effect. Therefore, the effect will be greater when H is replaced by D than when ^{79}Br is replaced by ^{81}Br. Suitable preparation and detection procedures must be available and a radioactive isotope must have an appropriate half-life for isotopically labeled materials to be used. This limits the range of atoms used in studies of the kinetic isotope effect considerably. Other than studies involving the isotopes of hydrogen, studies using ^{13}C or ^{14}C, ^{15}N, ^{34}S, ^{35}Cl or ^{37}Cl, and ^{79}Br and ^{80}Br are most common.

For a vibrating diatomic molecule A—B, the vibrational energy can be expressed as

$$E = h\nu(n + 1/2) \tag{3.48}$$

where h is Planck's constant, n is the vibrational quantum number, and ν is the frequency of the stretching vibration. For most diatomic molecules, the spacing between vibrational levels is on the order of 10 to 40 kJ/mol, and at room temperature, RT amounts to only about 2.5 kJ/mol. Therefore, practically all of the molecules will populate the lowest vibrational level ($n = 0$). Under these conditions or even at 0 K, the molecules will still have some vibrational energy (the zero-point vibrational energy),

$$E = (1/2)h\nu \tag{3.49}$$

If the vibration takes place with the molecule behaving as a harmonic oscillator, the frequency is given by

$$\nu = \frac{1}{2\pi}\left(\frac{f}{\mu}\right)^{1/2} \tag{3.50}$$

where f is the force constant for the bond and μ is the reduced mass, $m_A m_B/(m_A + m_B)$. The chemical bonds in A—B and A—B* (B and B* being different isotopes) are very nearly identical because electronic energies are essentially unaffected by the number of neutrons in the nuclei. The reduced mass is affected as we can easily illustrate. Consider the molecules H_2, HD, and HT (T=tritium, ^3H). For H_2,

$$\mu_{HH} = \frac{m_H m_H}{m_H + m_H} = \frac{m_H^2}{2m_H} = (1/2)m_H \tag{3.51}$$

Since $m_D \approx 2m_H$, for HD we have

$$\mu_{HD} = \frac{m_H m_D}{m_H + m_D} \approx \frac{m_H(2m_H)}{m_H + 2m_H} = (2/3)m_H \tag{3.52}$$

Similarly, for HT we find that $\mu \approx (3/4)m_H$ while for D_2 the result is $\mu \approx m_H$. Only in the case of the hydrogen isotopes is the relative mass effect this large.

The effect of reduced mass on the zero-point vibrational energy is easily seen. If we consider the molecules H—H and H—D in their lowest vibrational states, we find that the vibrational energy in terms of force constant, f, is

$$E = \frac{h}{4\pi}\left(\frac{f}{\mu}\right)^{1/2} \tag{3.53}$$

For the two molecules, the ratio of the energies is

$$\frac{E_{HH}}{E_{HD}} = \left(\frac{\mu_{HD}}{\mu_{HH}}\right)^{1/2} \approx \left(\frac{(2/3)m_H}{(1/2)m_H}\right)^{1/2} = (1.33)^{1/2} = 1.15 \tag{3.54}$$

This ratio of 1.15 is equal to that observed because the zero-point vibrational energies for H—H and H—D are 25.9 and 22.4 kJ/mol, respectively. Similarly, it can be shown that $E_{HH}/E_{DD} = 1.41$. Since $\mu_{HD} > \mu_{HH}$, it is found that $E_{HH} > E_{HD}$ and the zero-point vibrational energy is greater for the H—H bond than it is for the H—D bond.

Because H_2 already resides in a higher energy state than does HD, it requires correspondingly less energy to dissociate the H_2 molecule. Accordingly, a reaction that requires the dissociation of these molecules will take place more rapidly for H_2 than for HD.

For H or D atoms bound to another atom X which has a much larger mass,

$$\mu_{HX} = \frac{m_H m_X}{m_H + m_X} \approx m_H (\text{or } m_D) \tag{3.55}$$

Therefore, in the case where X is of a large mass,

$$\frac{E_{HX}}{E_{DX}} \approx \left(\frac{m_D}{m_H}\right)^{1/2} = 2^{1/2} = 1.41 \tag{3.56}$$

For O—H bonds, the vibrational absorption is at ~3600 cm^{-1}, while for O—D bonds it is at ~2600 cm^{-1}. Similarly, the vibrational absorption for C—H is at ~3000 cm^{-1}, but for D—C it is at ~2100 cm^{-1}. This suggests that the differences in the X—H bonds should give rise to a kinetic isotope effects when reactions occur at the X—H bonds.

Consider two reactant molecules that are identical except that one contains a different isotope in the reactive site. If the bonds in the reactant molecules to the two isotopic atoms are not broken in forming the transition state, the extent to which isotopic labelling affects the rate will be less than when the bond is completely broken. If the formation of the transition state does not alter the bond holding the isotopic atoms, there will be no isotope effect. However, if the bond becomes stronger to the different isotopes in the reactant molecules, there will be an *inverse* isotope effect. This results from the fact that as the bond becomes stronger in the transition state, the heavier isotope will give a transition state having a lower zero-point vibrational energy. Because this gives an overall lowering of the energy of the transition state above the reactant state, there will be a rate *increase* in the case of the *heavier* isotope.

Earlier, we described the reaction of chlorobenzene with amide ion to produce aniline. The mechanism involves the removal of H by NH_2^- to form the benzyne intermediate.

$$\text{(3.57)}$$

Therefore, the rate of the reaction should be subject to a kinetic isotope effect if deuterium replaces hydrogen in the 2-position. When chlorobenzene-2d is used, the ratio of the rate constants is $k_H/k_D = 5.5$. This large kinetic isotope effect indicates that breaking of the C—H bond occurs in the rate-determining step. As expected, the rate of breaking the X—H bond is higher than that for breaking the X—D bond.

To this point, we have presumed that bond-breaking occurs. Because quantum mechanically it is possible for barrier penetration to occur, tunneling must be considered as a reaction pathway. The transmission of a particle through a potential barrier is one of the basic models of quantum mechanics. We do not show the details of the solution here, but it can be shown that the *transparency* (transmission coefficient) of a rectangular barrier of height U and thickness x to particles of mass M having an energy E is given by

$$\mathbf{T} = \exp\left(-2x\frac{8\pi^2 M}{h^2}(U-E)^{1/2}\right) \qquad \text{(3.58)}$$

Therefore, the effects of several variables on barrier penetration are as follows:

1. The transparency decreases the higher the barrier, U.
2. The transparency increases the higher the energy of the particles, E.
3. The transparency increases the smaller the mass of the particles, M.
4. The transparency decreases as the thickness of the barrier, x, increases.
5. If $h = 0$, we have the classical limit where quantum conditions do not apply and $\mathbf{T} = 0$. That is, the particle can not pass the barrier because it has an energy lower than the barrier height.

It is seen that the smaller the mass of the particle, the greater the probability for barrier penetration. Likewise, the higher the energy of the particle, the greater the transmission coefficient. Both of these factors favor barrier penetration of H over that of D, so reactions involving tunneling also show a kinetic isotope effect which predicts the lighter isotopes react faster.

Although we have considered only the separation of diatomic molecules, the conclusions reached are still generally valid for more complex molecules. Bending vibrations are altered during a bond-breaking reaction, but because bending vibrations involve considerably lower energies than do stretching vibrations, they can be ignored in a qualitative approach. Therefore, breaking a bond in a polyatomic molecule is considered to be much like that in a diatomic molecule. There may also be other effects produced by isotopic substitution at positions other than the reactive site in the molecule. These effects are usually much smaller than primary isotope effects and are called *secondary* isotope effects. A very large number of reactions have been studied to use kinetic isotope effects to determine information about the transition state. For further details, the references at the end of the chapter should be consulted.

REFERENCES FOR FURTHER READING

Bernasconi, G. F., Ed. (1986) *Investigation of Rates and Mechanisms of Reactions Part I, Investigations of Rates and Mechanisms of Reactions*, Vol. VI in A. Weissberger, Ed., *Techniques of Chemistry*, 4th ed., Wiley, New York. Numerous chapters dealing with all aspects of kinetics in over 1000 pages.

Bernasconi, G. F., Ed. (1986) *Investigation of Rates and Mechanisms of Reactions Part II, Investigation of Elementary Reaction Steps in Solution and Fast Reaction Techniques*, Vol. VI in A. Weissberger, Ed., *Techniques of Chemistry*, 4th ed., Wiley, New York. This book deals with many aspects of reactions in solution and solvent effects.

Caldin, E. F. (1964) *Fast Reactions in Solution*, Blackwells, Oxford.

Cox, B. G. (1994) *Modern Liquid Phase Kinetics*, Oxford, New York, Chapters 2 and 5. A good introduction to the use of flow methods at an elementary level.

Mares, M., Palmer, D. A., Kelm, H. (1978) *Inorg. Chim. Acta, 27,* 153.

Melander, L., Saunders, W. H., Jr. (1980) *Reaction Rates of Isotopic Molecules*, Wiley, New York. A standard reference in the field of isotope effects.

Nicholas, J. (1976) *Chemical Kinetics: A Modern Survey of Gas Phase Reactions*, Halsted Press, New York. A good introduction to theory and practice in the study of gas phase reactions.

Roberts, J. D., Semenow, D. A., Simmons, H. E., and Carlsmith, L. A. (1956) *J. Am. Chem. Soc., 78,* 601.

Wentrup, C. (1986) Tracer Methods, in *Part I, Investigations of Rates and Mechanisms of Reactions*, Vol. VI in A. Weissberger, Ed., *Techiniques in Chemistry*, 4th ed., Wiley, New York.

PROBLEMS

1. For a reaction $aA + bB \longrightarrow$ Products, the initial rate varies with concentrations as follows.

$[A]_o$	$[B]_o$	R_i (M^{-1} sec^{-1})
0.0260	0.0320	0.000410
0.0170	0.0190	0.000159

Use the method of initial rates and determine the overall order of the reaction.

2. For the reaction

$$3\,O_3 + Br_2O \longrightarrow 3\,O_2 + 2\,BrO_2$$

decide some aspect of the mechanism that would have different outcomes if two isotopes were used. Write the question to be answered. Next, decide which species could be replaced by a different isotope and show how the mechanism could be elucidated by the use of a labeled compound.

3. For the reaction

$$N_2H_5^+ + HNO_2 \longrightarrow HN_3 + H^+ + 2\,H_2O$$

decide some aspect of the mechanism that would have different outcomes if two isotopes were used. Write the question to be answered. Next, decide which species could be replaced by a different isotope and show how the mechanism could be elucidated by the use of a labeled compound.

4. For a reaction $aA + bB \longrightarrow$ Products, the following data were obtained.

$[A]_o$	$[B]_o$	R_i (M^{-1} sec^{-1})
0.125	0.216	0.0386
0.186	0.216	0.0568
0.125	0.144	0.0176

Use the logarithmic method to determine the rate law for the reaction.

5. For the reaction

$$[Co(NH_3)_5ONO]^{2+} \longrightarrow [Co(NH_3)_5NO_2]^{2+}$$

carried out at 30°C, the rate constant varies with pressure as follows (Mares, M, Palmer, D. A., Kelm, M. *Inorg. Chim. Acta*, **1978**, 27, 153.) Use these data to determine the volume of activation for the reaction.

P, bar	$10^5 k$, sec^{-1}	P, bar	$10^5 k$, sec^{-1}
1	13.7 ± 0.3	1000	20.2 ± 0.5
250	15.9 ± 0.3	1500	21.5 ± 0.4
500	17.5 ± 0.5	2000	23.9
750	18.4 ± 0.4	2500	27.8

6. The reaction

$$2 \; 2\text{-}C_5H_{10} \longrightarrow 2 - C_4H_8 + 3\text{-}C_6H_{12}$$

is catalyzed by $C_5H_5NMo(NO)_2Cl_2$ (Hughes, W. B. *J. Am. Chem. Soc.* **1970**, *92*, 532). When the catalyst concentration is 2.08×10^{-3} **M**, the rate of 2-C_5H_{10} loss is 0.73×10^2 **M** min^{-1} and when the catalyst concentration is 4.16×10^{-3} **M**, the rate is 1.30×10^2 **M** min^{-1}. Use these data and the logarithmic method to determine the order with respect to the catalyst.

chapter 4

REACTIONS IN THE GAS PHASE

In the previous chapters, we have considered reactions on an empirical basis in terms of several concentration-time relationships. Our intuition tells us that while the overall reaction may be described in this way, on a molecular level, individual reacting "units" must on some microscopic scale collide or make contact in some way. These units—molecules, ions, atoms, radicals, and electrons—must be involved in some simplest step at the instant of reaction. These steps through which individual units pass are called *elementary* reactions. The sequence of these elementary reactions constitutes the mechanism of the reaction. For reactions in the gas phase, molecular collisions constitute the vehicle for energy transfer, and our description of gas phase reactions begins with a kinetic theory approach to collisions of gaseous molecules. In simplest terms, the two requirements for a reaction to occur are (1) a collision must occur and (2) the molecules must possess sufficient energy to cause a reaction to occur. It will be shown that this treatment is not sufficient, but it is the starting point for the theory.

4.1   COLLISION THEORY

In the simplest interpretation, the rate constant contains information related to factors determining the rate of a reaction. For example, because

$$k = Ae^{-E/RT} \tag{4.1}$$

E is related to the energy barrier over which the reactants must pass. For molecules that are colliding, $e^{-E/RT}$ is related to the number of molecular collisions that have the required energy to induce reaction. The pre-exponential factor, A, is related to the frequency of collisions. Therefore, we can describe the reaction rate as

$$\text{Rate} = \text{Collision frequency} \times \left(\begin{array}{c} \text{Fraction of collisions with} \\ \text{at least the threshold energy} \end{array} \right)$$

$$\text{Rate} = Z_{AB} \times F \tag{4.2}$$

where Z_{AB} is the frequency of collisions between molecules of A and B, and F is the fraction of those collisions having sufficient energy to cause reaction.

The collision frequency between two different types of molecules can be calculated using the kinetic theory of gases. In this treatment, we will consider the molecules of B as being stationary and A molecules moving through them. If we imagine a molecule of A moving through space, collisions will occur with molecules of B whose centers lie within a cylinder of length v_{AB} and radius $r_A + r_B$ where v_{AB} is the average relative velocity of A and B, and $r_A + r_B$ is the sum of the radii of molecules A and B. This situation is shown in Figure 4.1.

We can call the cross-sectional area of the cylinder, $\pi(r_A + r_B)^2$, the *collisional cross section*, σ_{AB}. In one second, the molecule of A travels a distance of v_{AB} (v_{AB} is the average molecular velocity of A relative to B) and it will collide with all B molecules that have centers within the cylinder. Therefore, the number of collisions per second is the number of B molecules/cm³ multiplied by the volume of the cylinder.

$$Z_A = v_{AB}\sigma_{AB}C_B \qquad (4.3)$$

Although A does not continue in a straight line after colliding with B, the calculated collision frequency will still be correct as long as there is no gradient in concentration of B within the container and the velocity of A is constant. The result above is for a single molecule of A. To obtain the total number of colli-

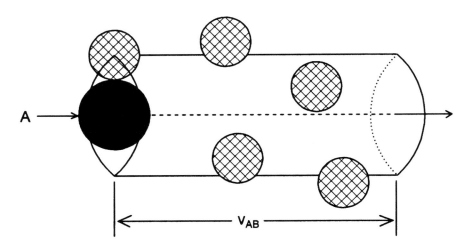

Figure 4.1   Model used for calculating collision frequency.

sions between molecules of A and B, Z_{AB}, we must multiply the results above by C_A, the number of molecules of A per cm^3.

$$Z_{AB} = v_{AB}\sigma_{AB}C_AC_B \tag{4.4}$$

We have considered molecules of B to be stationary (velocity = 0) so the relative velocity v_{AB} is just the average velocity of A,

$$v_A = \left(\frac{8kT}{m\pi}\right)^{1/2} \tag{4.5}$$

where T is the temperature (K), **k** is Boltzmann's constant, and m is the mass of A. If we represent the reduced mass of a pair of molecules A and B as μ, then

$$\frac{1}{\mu} = \frac{1}{m_A} + \frac{1}{m_B}$$

or, in the more familiar form,

$$\mu = \frac{m_A m_B}{m_A + m_B} \tag{4.6}$$

The relative velocity of A and B can now be written as

$$v_{AB} = \left(\frac{8kT}{\pi\mu}\right)^{1/2} \tag{4.7}$$

Therefore,

$$Z_{AB} = \left(\frac{8kT}{\pi\mu}\right)^{1/2}\sigma_{AB}C_AC_B \tag{4.8}$$

Frequently, the collision diameter, $d = (d_A + d_B)/2$, is used and the concentrations are written in terms of numbers of molecules/cm^3, n_A and n_B, per unit volume. Then,

$$Z_{AB} = d^2\pi\left(\frac{8kT}{\pi\mu}\right)^{1/2}\frac{n_An_B}{V^2} \tag{4.9}$$

If we consider 1 cm^3 of gaseous H_2 at 1 atm and 300 K, using a collision diameter of 0.21 nm (2.1×10^{-8} cm), we obtain a collision frequency of about 1.8 $\times 10^{29}$ collisions per second per cm^3. Therefore, at this collision frequency, if every collision led to a reaction, 1 mole of a gas could react in about 6.02×10^{23} molecules reacting/1.8×10^{29} molecules reacting/sec or about 3.3×10^{-6} sec. Because most gaseous reactions occurring between colliding molecules do not take place on this time scale, other factors than just the collision frequency must be considered. We must now consider these other factors.

One factor that has been ignored to this point is that while a collision frequency can be calculated, the collision between the molecules must occur with sufficient energy for the reaction to occur. As we have previously seen, that minimum energy is the activation energy. Figure 4.2 shows a Maxwell-Boltzmann distribution of energies of gaseous molecules. If the minimum energy is the activation energy, E_a, the fraction (F) of the molecules possessing that energy or greater (given by the shaded area under the curve) is given by

$$F(E) = \frac{\displaystyle\int_E^\infty e^{-E/RT} dE}{RT} \tag{4.10}$$

or

$$F = e^{-E_a/RT} \tag{4.11}$$

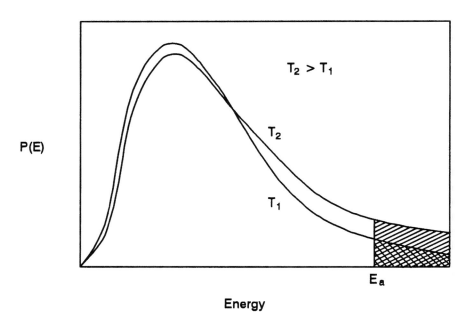

Figure 4.2   Distribution of molecular energies.

Even if the activation energy is small, the fraction of the molecules having a collision energy leading to reaction will be a very small fraction of the total number of collisions. The reaction rate should be given by

$$\text{Rate} = \text{Collision frequency} \times \text{fraction of collisions with } E > E_a$$

$$\text{Rate} = d^2\pi\left(\frac{8kT}{\pi\mu}\right)^{1/2} C_A C_B e^{-E/RT} \tag{4.12}$$

It should be noted here that the collision theory of reaction rates predicts that the pre-exponential factor is not independent of temperature, but rather depends on $T^{1/2}$. This occurs because the average *kinetic energy* of an ideal gas is directly proportional to T but the velocity, which determines collision frequency, is proportional to $T^{1/2}$. Over a narrow range of temperature, this dependence on temperature is not observed. The fact is that a rather slight increase in temperature changes the average molecular velocity only very slightly because of a factor of $(T_2/T_1)^{1/2}$. However, the area under the curve at energies higher than E is increased slightly (Figure 4.2). Therefore, the dominant temperature effect occurs in the $e^{-E/RT}$ factor where the fraction of molecules having $E > E_a$ is calculated. As was illustrated in Chapter 2, an increase in temperature of $10°$ can double or triple the rate of a reaction.

When reaction rates calculated using collision theory are compared to the experimental rates, the agreement is usually poor. In some cases, the agreement is within a factor of two or three, but in other cases the calculated and experimental rates differ by 10^5 to 10^7. The discrepancy is usually explained in terms of *effective* collisions, which are only a fraction of the *total* collisions owing to steric requirements. The idea here is that in order for molecules to react, (1) a collision must occur, (2) the collision energy must be sufficient, and (3) the molecules must have a proper orientation. A steric factor, P, is defined as

$$P = \frac{\sigma_{obs}}{\sigma_{calc}} \tag{4.13}$$

This steric factor can be regarded as an orientation factor, but it can also be interpreted in terms of the entropy change involved in forming the transition state.

4.2   THE POTENTIAL ENERGY SURFACE

It is reasonable to assume that the elementary reaction

$$Cl + H\text{---}H \longrightarrow [Cl\cdots H\cdots H]^* \longrightarrow HCl + H \tag{4.14}$$

passes through a linear Cl—H—H transition state. That the transition state is linear in this case follows from the fact that to form a bent transition state would bring the terminal atoms closer together increasing repulsion. Now, to relate the energy of this system to the bond distances is the problem. While we might consider this problem in a number of ways, the simplest approach is to extend the relationship used for a diatomic molecule to include a second bond.

The bond energy of a diatomic molecule as a function of bond length is shown in Figure 4.3. One equation that gives the kind of relationship shown is the Morse equation,

$$E = D_e [e^{-2\beta(r-r_o)} - 2e^{-\beta(r-r_o)}]$$ (4.15)

where r is the internuclear distance, r_o is the equilibrium internuclear distance, D_e is the bond dissociation energy, and β is a constant. Attraction increases as the atoms get closer together, but at distances smaller than r_o, repulsion increases and becomes dominant at very short internuclear distances.

For a linear triatomic transition state, it is assumed that a second potential energy curve results so that the total energy is a function of two bond distances. Therefore, a diagram can be constructed which shows energy on one axis, one bond distance on another, and the other bond distance on the third axis, generating a surface. If we suppose the reaction

$$AB + C \longrightarrow BC + A$$ (4.16)

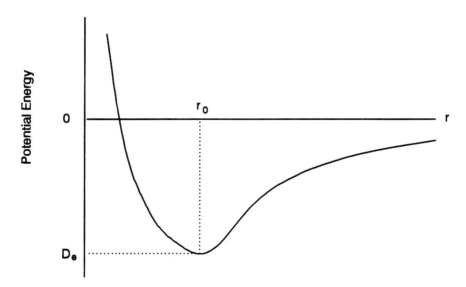

Figure 4.3   Potential energy curve for a diatomic molecule.

takes place with the formation of a linear transition state A $\cdots$ B $\cdots$ C, the resulting three-dimensional surface is like that shown in Figure 4.4. In order to go from AB + C to the products BC + A, it is not necessary to go over the high energy region (large A $\cdots$ B $\cdots$ C distances). Instead, the reaction proceeds along a path where the energy rises less steeply along a "valley." Along that path, the energy barrier is lower, being similar to a pass over a mountain range. Such a path passes over a highest point sometimes referred to as a "saddle" point. The path representing the changes in configuration as the reaction takes place is called the *reaction coordinate*.

Since electronic energy levels for molecules differ by perhaps 200 to 400 kJ/mol and the motion of electrons is rapid compared to the motion of nuclei within the molecules, it is possible to determine the energy as if the nuclei are at rest (the Born-Oppenheimer approximation). The assumption is made that the coulombic and exchange energies are related by an approximately constant ratio (the exchange energy is approximately 15% of the coulombic energy). For a diatomic molecule, the energy can be written as

$$E = \frac{Q \pm J}{1 + S^2} \tag{4.17}$$

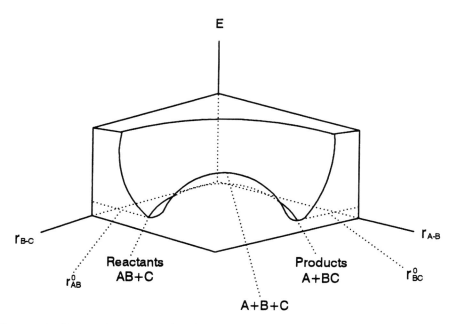

Figure 4.4   Potential energy surface for AB + C $\longrightarrow$ A + BC.

where Q, J, and S are the coulombic, exchange, and overlap integrals, respectively. For a triatomic molecule, the energy can be written as

$$E = Q_A + Q_B + Q_C \pm \{(1/2)[(J_A - J_B)^2 + (J_B - J_C)^2 + (J_A - J_C)^2]\}^{1/2} \quad \textbf{(4.18)}$$

where Q_A is the coulombic term for B – C, J_A is the exchange term for B – C; Q_B and J_B are the coulombic and exchange terms for A – C, and Q_C and J_C are the coulombic and exchange terms for A – B. This method, developed by London, does not reproduce known energies accurately and it results in a "basin" at the top of the saddle point. While more exact calculations based on the variation method and semi-empirical procedures provide results which are in qualitative agreement with experimental results, especially for simple molecules, the details of these methods will not be presented here.

Another facet of the potential energy barrier to reaction is that of quantum mechanical tunneling. Classically, an object must have an energy at least equivalent to the height of a barrier in order to pass over it. Quantum mechanically, it is possible for a particle to pass through a barrier even though the particle has an energy that is less than the height of the barrier. In the particle in the one-dimensional box model, the walls of the box are made infinitely high to prevent the particle from "leaking" from the box. The *transparency* of a barrier is determined by the height and thickness of the barrier and the mass and energy of the particle. For a given barrier, the transparency decreases as the mass of the particle increases so that tunneling is greater for light atoms (e.g., H, H$^+$, etc.). However, the transparency increases as the energy approaches the barrier height (see Section 3.9).

A potential energy surface such as that shown in Figure 4.4 is symmetrical, indicating that the diatomic AB is very similar to BC. The potential energy curves indicate that they have similar bond energies. In a more general case, the reactant and product molecules will have different bond energies so the potential energy surface will not be as nearly symmetrical, as shown in Figure 4.5. In the case shown, the product molecule lies at a lower energy than the reactant showing that the reaction is exothermic.

An alternative method of showing a potential energy surface is based on the same principle as that used to prepare a topographical map. On a topographical map, lines connect points of equal altitude creating contours having given altitudes. Where the contour lines are closely spaced, the altitude changes abruptly, and where the contour lines are widely separated, the land is essentially flat. Slices through the surface at specific constant energies provide the view shown in Figure 4.6. This case corresponds to the reaction in which the molecules BC and AB have similar bond energies. For the case where BC and AB have greatly different bond energies, the surface will have one of the "valleys" as being deeper and have steeper walls, as indicated by more closely spaced contour lines. This situation is shown in Figure 4.7.

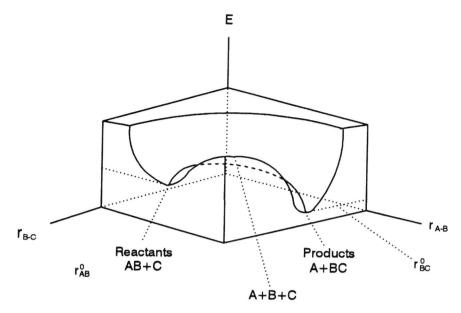

Figure 4.5   Potential energy surface for an exothermic reaction.

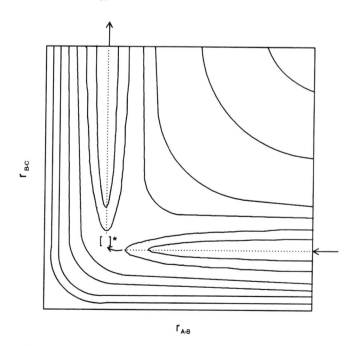

Figure 4.6   Contours of constant potential energy for $A \cdots B \cdots C$.

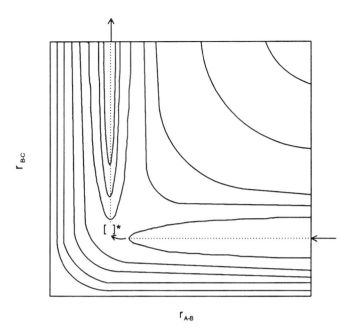

Figure 4.7 Contours of constant potential energy for A $\cdots$ B $\cdots$ C.

4.3 TRANSITION STATE THEORY

A collision theory of even gas phase reactions is not satisfactory, and the problems with the steric factor that we described earlier make this approach more empirical and qualitative than we would like. Transition state theory, developed largely by Henry Eyring, takes a somewhat different approach. We have already considered the potential energy surfaces which provide a graphical energy model for chemical reactions. Transition state theory (or activated complex theory) refers to the details of how reactions become products. For a reaction such as

$$AB + C \longrightarrow BC + A$$

it is assumed that there is a variation in potential energy which is related to atomic coordinates by a potential energy function. The term *phase space* is applied to the coordinate and momentum space for the system. In order for a reaction to occur, the transition state must pass through some critical configuration in this space. Because of the nature of the potential function used to express the energy of the system as a function of atomic positions, the system energy possesses a saddle point. This saddle point lies lower in energy than the fully dissociated arrangement, A + B + C, or the highly "compressed" arrangement, A – B – C. The nature of such a surface is shown in Figure 4.4.

The essential feature of transition state theory is that there is a "concentration" of the species at the saddle point, the activated complex, that is in equilibrium with reactants and products. The Boltzmann Distribution Law governs the concentration of that activated complex, and the rate of reaction is proportional to that concentration. Since the concentration of activated complex is small because of its energy being higher than that of the reactants, this critical configuration represents the regulator of the rate of flow of reactants to products.

The concentration of the activated complex is not the only factor involved because the *frequency* of its dissociation into products comes into play. Therefore, the rate can be expressed as

$$\text{Rate} = \left(\begin{array}{c}\text{Activated complex}\\\text{concentration}\end{array}\right)\left(\begin{array}{c}\text{Decomposition frequency}\\\text{of the activated complex}\end{array}\right)$$

In order for the activated complex to separate into products, one bond (the one being broken) must acquire sufficient vibrational energy to separate. When it does separate, one of the $3N - 6$ vibrational degrees of freedom is lost and is transformed into translational degrees of freedom of the products. Central to the ideal of transition state theory is the assumption that the activated complex is in equilibrium with the reactants. Thus,

$$A + B \rightleftharpoons [AB]^{\ddagger} \longrightarrow \text{Products} \tag{4.19}$$

For the formation of the activated complex, $[AB]^{\ddagger}$, the equilibrium constant is

$$K^{\ddagger} = \frac{[AB]^{\ddagger}}{[A][B]} \tag{4.20}$$

from which we find

$$[AB]^{\ddagger} = K^{\ddagger}[A][B] \tag{4.21}$$

Since the reaction rate is written as

$$\text{Rate} = \left(\begin{array}{c}\text{Activated complex}\\\text{concentration}\end{array}\right)\left(\begin{array}{c}\text{Decomposition frequency}\\\text{of the activated complex}\end{array}\right)$$

we can now write

$$\text{Rate} = [AB]^{\ddagger} \times (\text{frequency}) = (\text{frequency})K^{\ddagger}[A][B] \tag{4.22}$$

As we have seen previously (see Eq. 2.109)

$$K^{\ddagger} = \exp(-\Delta G^{\ddagger} / RT)$$

and we know that

$$\Delta G^{\ddagger} = \Delta H^{\ddagger} - T\Delta S^{\ddagger}$$

therefore,

$$K^{\ddagger} = \exp[-(\Delta H^{\ddagger} - T\Delta S^{\ddagger})/RT] = \exp(-\Delta H^{\ddagger}/RT) \cdot \exp(\Delta S^{\ddagger}/R) \qquad \textbf{(4.23)}$$

Substituting for $K^{\ddagger}$ in Eq. (4.22) yields

$$\text{Rate} = (\text{frequency})[A][B] \; \exp(-\Delta H^{\ddagger}/RT) \cdot \exp(\Delta S^{\ddagger}/R) \qquad \textbf{(4.24)}$$

The frequency of decomposition of the activated complex must now be addressed.

If we consider the vibration of the activated complex at the top of the potential barrier, it is instructive to recall that the classical high temperature limit for a vibrational mode is

$$E_{vib} = kT \text{ (per molecule)} \qquad \textbf{(4.25)}$$

where k is Boltzmann's constant and T is the temperature (K). Because $k = R/N$ (N is Avogadro's number),

$$E_{vib} = RT \text{ (per mole)} \qquad \textbf{(4.26)}$$

It should be remembered that for each degree of translational freedom, the energy is $kT/2$, which is $RT/2$ per mole. If we now assume that the frequency of activated complex decomposition is the frequency of the vibration being lost due to breaking the bond,

$$E = h\nu = kT$$

or

$$\nu = kT/h \qquad \textbf{(4.27)}$$

Since the reaction rate in terms of the activated complex concentration is

$$\text{Rate} = k[AB]^{\ddagger} \qquad \textbf{(4.28)}$$

The rate of the reaction can now be written as

$$\text{Rate} = \nu K^{\ddagger}[A][B]$$

and substitution for the frequency gives

$$\text{Rate} = \frac{kT}{h} K^{\ddagger}[A][B] \qquad \textbf{(4.29)}$$

and the rate constant is

$$k = \frac{kT}{h} K^{\ddagger} = \frac{kT}{h} \exp(-\Delta G^{\ddagger}/RT) \tag{4.30}$$

A somewhat more elegant approach to deriving an expression for the rate of passage over the potential energy barrier is based on statistical mechanics. According to this procedure, it is assumed that there is a certain distance, d, at the top of the barrier which must be the distance where the activated complex exists. It is within this distance that a vibrational mode of the complex is transformed into translational motion of the products. The rate of passage of the activated complex through distance d is related to the molecular velocity in one direction. If the mass of the activated complex is $m^{\ddagger}$, the velocity is

$$v = (2kT/\pi m^{\ddagger})^{1/2} \tag{4.31}$$

Therefore, the time required for the activated complex to pass through distance d is

$$\frac{d}{v} = d\left(\frac{m^{\ddagger}\pi}{2kT}\right)^{1/2} \tag{4.32}$$

The number of complexes crossing the potential barrier through distance d per unit time is

$$\frac{d[\ddagger]}{dt} = \frac{[\ddagger]/2}{d(\pi m^{\ddagger}/2kT)^{1/2}} = \frac{[\ddagger]}{d}\left(\frac{kT}{2\pi m^{\ddagger}}\right)^{1/2} \tag{4.33}$$

Note that we are using [ ]‡ to represent the activated complex and [$\ddagger$] to represent the *concentration* of the activated complex. Now, the concentration of the activated complex, [$\ddagger$], is to be evaluated. If the difference between the zero-point energies of the reactants and the complex is represented as $E_o^{\ddagger}$, the equilibrium constant for formation of the activated complex is

$$K^{\ddagger} = \frac{Q^{\ddagger}}{Q_A Q_B} \exp(-E_o^{\ddagger}/RT) \tag{4.34}$$

where Q_A, Q_B, and $Q^{\ddagger}$ are the partition functions of reactants A and B and the activated complex, respectively. If the vibrational mode of the activated complex is factored out of $Q^{\ddagger}$, we can write

$$Q^{\ddagger} = Q^{\ddagger\prime} \times q_v^{\ddagger} \tag{4.35}$$

where $q_v^{\ddagger}$ is the vibrational mode of the bond being broken. Now we can approximate the vibrational mode as

$$q_v^{\ddagger} = \frac{1}{(1 - \exp(-hv/kT))} = \frac{kT}{hv^{\ddagger}} \tag{4.36}$$

and the equilibrium constant $K^{\ddagger}$ is

$$K^{\ddagger} = \frac{kT}{hv^{\ddagger}} \cdot \frac{Q^{\ddagger\prime}}{Q_A Q_B} \exp(-E_o^{\ddagger}/RT) \tag{4.37}$$

which is of the same form found earlier with the rate constant being given by

$$k = \frac{kT}{h} \cdot \frac{Q^{\ddagger\prime}}{Q_A Q_B} \exp(-E_o^{\ddagger}/RT) \tag{4.38}$$

The resemblance of this equation to the Arrhenius equation is obvious when the pre-exponential factor includes the frequency factor and the equilibrium constant in terms of partition functions. This expression for k is similar to that obtained from collision theory.

The approximate rate constant, k^a, can be calculated from probabilities that the reactants in the distribution of quantum states will collide and react based on collision frequency. That constant is greater than the actual measured rate constant, k. One approach to improving transition state theory with respect to calculating the rate constant is to alter the configuration of the transition state used in the energy calculations in order to effect a change in k^a. In fact, the calculations are performed in such a way that the calculated rate constant is a minimum and thereby approaches the observed k. Just as energy minimization is accomplished by the variation method, this procedure is referred to in this connection as *variational* transition state theory.

Since

$$\Delta G^{\ddagger} = -RT \ln K^{\ddagger}$$

this procedure amounts to configuration optimization to minimize $K^{\ddagger}$ or maximize $\Delta G^{\ddagger}$. In practice, a series of transition states is considered and the calculations are performed to obtain the desired minimization. It is of some consequence to choose the reaction path with respect to the energy surface. Generally, the path chosen is the path of steepest descent on either side of the saddle point. This path represents the path of minimum energy. While the details will not be presented here, the rate constant is now treated as a function of a coordinate-related parameter, z, so that

$$k(z)^{VT} = \frac{kT}{h} \frac{Q^{VT}}{Q_A Q_B} e^{-E(z)/RT} \tag{4.39}$$

where z is the path-related parameter. This expression can also be written as

$$k(z)^{VT} = \frac{kT}{h} K^{\ddagger} e^{-\Delta G^{\ddagger}(z)/RT} \tag{4.40}$$

The value of $k(z)^{VT}$ is now minimized with respect to z. Accordingly, the rate constant is minimized with respect to a parameter related to configuration of the transition state in the same way that energy is minimized with respect to variables in a trial wave function. Although this topic will not be described further here, details have been published in several places (e.g., Truhlar, 1980).

4.4 UNIMOLECULAR DECOMPOSITION OF GASES

The collision theory of gaseous reactions requires *two* molecules to collide, suggesting that such reactions should be second-order. Many decompositions (e.g., N_2O_5) are first-order at sufficiently high pressures of the gas. However, some such reactions do appear to be second-order at low gas pressure. In 1922, Lindemann proposed an explanation of these observations.

Molecules transfer energy as a result of molecular collisions. Therefore, translational energy can be transferred to one molecule by another, raising the translational and vibrational energy of the second molecule. The *collisional* activation of molecules can thus be accomplished. However, the activated molecule need not react immediately, and, in fact, it may become deactivated by undergoing subsequent collisions before it reacts. For reaction to occur, the activated molecule, which has increased vibrational energy, must have some bond activated to the point where bond rupture occurs.

The elementary reactions by which A is converted into products can be shown as

$$A + A \underset{k_{-1}}{\overset{k_1}{\rightleftharpoons}} A + A^* \tag{4.41}$$

$$A^* \xrightarrow{k_2} \text{Products} \tag{4.42}$$

In this scheme, A^* is the activated molecule of A. While the process producing A^* is bimolecular, the decomposition of A^* is unimolecular. The rate of the reaction can be written as

$$-\frac{d[A]}{dt} = k_1[A]^2 - k_{-1}[A][A^*] \tag{4.43}$$

Because A* is an activated molecule, a reactive intermediate, we can use the steady-state approximation (see Section 2.4). For a steady state, the rate of formation of A* is assumed to be equal to its rate of decomposition. Therefore,

$$\frac{d[A^*]}{dt} = 0 = k_1[A]^2 - k_{-1}[A][A^*] - k_2[A^*] \tag{4.44}$$

The first term on the right-hand side of the equation represents the rate of activation of A, while the second and third terms represent deactivation and decomposition of A*, respectively. Solving Eq. (4.44) for [A*], we obtain

$$[A^*] = \frac{k_1[A]^2}{k_{-1}[A] + k_2} \tag{4.45}$$

Substituting this result in Eq. (4.43), we obtain

$$-\frac{d[A]}{dt} = \frac{k_1 k_2 [A]^2}{k_{-1}[A] + k_2} \tag{4.46}$$

At high pressures, the number of A molecules per unit volume is large, and deactivation of A* can occur by frequent collisions with molecules of A. Under these conditions, the rate of deactivation of A* will be large compared to the rate of decomposition. Therefore, $k_{-1}[A] \gg k_2$ and

$$-\frac{d[A]}{dt} = \frac{k_1 k_2 [A]^2}{k_{-1}[A] + k_2} \approx \frac{k_1 k_2 [A]^2}{k_{-1}[A]} = k[A] \tag{4.47}$$

where $k = k_1 k_2 / k_{-1}$. Therefore, at relatively high pressure where [A] is high, the reaction appears to be unimolecular (first-order) in [A].

At low pressures of A, the rate of decomposition of A* is greater than the rate of its deactivation by collision with A because there are fewer molecules of A available. Under these conditions, the increase in vibrational energy can cause bond rupture and decomposition. Therefore, in this case, $k_2 \gg k_{-1}[A]$ and

$$-\frac{d[A]}{dt} = \frac{k_1 k_2 [A]^2}{k_2} = k_1[A]^2 \tag{4.48}$$

This equation shows that at low pressures of the reacting gas, the reaction should be bimolecular (second-order). Thus, the observed bimolecular dependence at low pressure and the unimolecular dependence at high pressure are predicted by a mechanism involving activation of molecules by collision.

The activation of reactant molecules by collision was described above. However, this is not the only vehicle for molecular activation. It is possible for a non-reactant gas to cause activation of molecules of the reactant. If we represent such a species by M, the processes of activation, deactivation, and product production are given by

$$A + M \underset{k_{-1}}{\overset{k_1}{\rightleftharpoons}} A^* + M \tag{4.49}$$

$$A^* \xrightarrow{k_2} \text{Products} \tag{4.50}$$

Therefore, the rate of disappearance of A can be written as

$$-\frac{d[A]}{dt} = k_1[A][M] - k_{-1}[A^*][M] \tag{4.51}$$

The change in concentration of A^* with time is given by

$$\frac{d[A^*]}{dt} = k_1[A][M] - k_{-1}[A^*][M] - k_2[A^*] = 0 \tag{4.52}$$

Therefore,

$$0 = k_1[A][M] - (k_{-1}[M] + k_2)[A^*]$$

so that

$$[A^*] = \frac{k_1[A][M]}{k_{-1}[M] + k_2} \tag{4.53}$$

Substituting for $[A^*]$ in Eq. (4.51) gives

$$-\frac{d[A]}{dt} = k_1[A][M] - k_{-1}[M]\frac{k_1[A][M]}{k_{-1}[M] + k_2} \tag{4.54}$$

Factoring out the quantity $k_1[A][M]$ on the right-hand side gives

$$-\frac{d[A]}{dt} = k_1[A][M]\left(1 - \frac{k_{-1}[M]}{k_{-1}[M] + k_2}\right) \tag{4.55}$$

The quantity inside the parentheses can be made into a single fraction since

$$-\frac{d[A]}{dt} = k_1[A][M]\left(\frac{k_{-1}[M] + k_2}{k_{-1}[M] + k_2} - \frac{k_{-1}[M]}{k_{-1}[M] + k_2}\right) \tag{4.56}$$

This equation can now be written as

$$-\frac{d[A]}{dt} = k_1[A][M]\left(\frac{k_{-1}[M] + k_2 - k_{-1}[M]}{k_{-1}[M] + k_2}\right) \tag{4.57}$$

which simplifies to

$$-\frac{d[A]}{dt} = \frac{k_1 k_2 [M][A]}{k_{-1}[M] + k_2} \tag{4.58}$$

The results obtained by considering activation by a third body must now be compared to those described above for activation by collision of reactant molecules. At high pressure, the rate of deactivation by collisions with M is greater than the rate of reaction so $k_{-1}[M] \gg k_2$ and

$$-\frac{d[A]}{dt} \approx \frac{k_1 k_2 [A][M]}{k_{-1}[M]} \approx k'[A] \tag{4.59}$$

and the reaction follows a first-order rate law. At low pressure, the concentration of M is low so $k_{-1}[M] \ll k_2$ and

$$-\frac{d[A]}{dt} \approx \frac{k_1 k_2 [A][M]}{k_2} \approx k_1[A][M] \tag{4.60}$$

and the reaction is first-order in A and first-order in M. If the species M is simply another molecule of reactant, this equation becomes

$$-\frac{d[A]}{dt} = k_1[A]^2 \tag{4.61}$$

which is the second-order expression found earlier. These results are in accord with experience for the unimolecular decomposition of a large number of gaseous compounds.

Ozone decomposes by a mechanism which appears to be somewhat different from that described above, but it provides a rather simple application of the steady state approximation. The reaction is

$$2O_3(g) \longrightarrow 3O_2(g) \tag{4.62}$$

for which the observed rate law is

$$-\frac{d[O_3]}{dt} = \frac{k[O_3]^2}{[O_2]} \tag{4.63}$$

Therefore, the reaction is second-order in ozone but the reaction is inhibited by O_2. This reaction is believed to involve a third body (an inert molecule or particle) in the steps,

$$O_3(g) + M \underset{k_{-1}}{\overset{k_1}{\rightleftharpoons}} O_2(g) + O(g) + M \quad \text{(fast)} \tag{4.64}$$

$$O(g) + O_3(g) \xrightarrow{k_2} 2O_2(g) \quad \text{(slow)} \tag{4.65}$$

The rate constant for the second reaction is much smaller than that for the first so that the second reaction is rate-determining.

$$-\frac{d[O_3]}{dt} = k_2[O][O_3] \tag{4.66}$$

The rate of [O] formation is given by

$$\frac{d[O]}{dt} = k_1[M][O_3] \tag{4.67}$$

and the rate of its consumption is

$$-\frac{d[O]}{dt} = k_{-1}[M][O][O_2] \tag{4.68}$$

Therefore, applying the steady-state assumption,

$$k_1[M][O_3] = k_{-1}[M][O][O_2] \tag{4.69}$$

Solving for [O], we obtain

$$[O] = \frac{k_1[O_3]}{k_{-1}[O_2]} \tag{4.70}$$

which when substituted in Eq. (4.66) gives

$$-\frac{d[O_3]}{dt} = k_2[O_3] \cdot \frac{k_1[O_3]}{k_{-1}[O_2]} = k\frac{[O_3]^2}{[O_2]} \tag{4.71}$$

This is the form of the observed rate law when $k = k_1 k_2 / k_{-1}$

The approach of Lindemann is based on collisional activation of molecules as a result of energy transfer. C. N. Hinshelwood (Nobel Prize, 1956) extended this approach to include changes in vibrational energies, which can be distributed internally to supply sufficient energy to the bond being broken. This approach provided a better fit to observed kinetics in the region of low pressure.

In the late 1920s, O. K. Rice and H. C. Ramsperger, as well as L. S. Kassel, developed an approach (the RRK theory) to unimolecular decomposition reactions, which is based on statistically treating the molecules as coupled oscillators. In this way, energy is presumed to be distributed about the energized molecule until it vibrates in a way that results in bond rupture. In this treatment, it is assumed that the amount of energy, E*, must be localized in the bond being broken and that the probability of this happening is given by

$$P = \left(\frac{E - E^*}{E}\right)^{N-1} \tag{4.72}$$

where N is the number of vibrational modes (3N – 5 for linear molecules and 3N – 6 for nonlinear molecules). The rate constant is therefore presumed to be given by

$$k = \left(\frac{E - E^*}{E}\right)^{N-1} \tag{4.73}$$

It can then be shown that at high pressure

$$k = \exp(-E^* / kT) \tag{4.74}$$

A later modification of this theory developed by R. A. Marcus (1952) (Nobel Prize, 1992) resulted in the so-called RRKM theory. In this case, the mechanism is assumed to consist of the following steps:

$$A + M \underset{k_{-1}}{\overset{k_1}{\rightleftharpoons}} A^* + M \tag{4.75}$$

$$A^* \xrightarrow{k_2} A^\ddagger \tag{4.76}$$

$$A^\ddagger \xrightarrow{k_3} \text{Products} \tag{4.77}$$

The essential idea is that the *activated molecule*, A*, becomes the *activated complex*, A‡, which then leads to product formation. This is presumed to occur when the energy at the reactive site becomes as large as E_a, the activation energy. The rate at which A^* is transformed into $A^{\ddagger}$ depends on the number of degrees of vibrational freedom. Therefore, the theory is concerned with the treatment of the vibrational frequencies of A^* and $A^{\ddagger}$ in the calculations.

From the processes above, we can write

$$-\frac{d[A]}{dt} = k_1[A][M] - k_{-1}[A^*][M] \qquad (4.78)$$

Applying the steady-state approximation to A^* gives

$$\frac{d[A^*]}{dt} = k_1[A][M] - k_{-1}[A^*][M] - k_2[A^*] = 0 \qquad (4.79)$$

Therefore, the concentration of A^* is

$$[A^*] = \frac{k_1[A][M]}{k_2 + k_{-1}[M]} \qquad (4.80)$$

From Eq. (4.79), we find that

$$k_{-1}[A^*][M] = k_1[A][M] - k_2[A^*] \qquad (4.81)$$

Therefore, Eq. (4.78) can be written as

$$-\frac{d[A]}{dt} = k_1[A][M] - k_{-1}[A^*][M] = k_1[A][M] - (k_1[A][M] - k_2[A^*])$$

or

$$-\frac{d[A]}{dt} = k_2[A^*] \qquad (4.82)$$

Now, substituting for $[A^*]$ gives

$$-\frac{d[A]}{dt} = \frac{k_2 k_1[A][M]}{k_2 + k_{-1}[M]} = \left(\frac{k_1 k_2[M]}{k_2 + k_{-1}[M]}\right)[A] \qquad (4.83)$$

We can now consider the quantity in brackets as the rate constant for the formation of product or disappearance of A. If $k_{-1}[M] \gg k_2$, the equation reduces to

$$-\frac{d[A]}{dt} = k'[A] \qquad (4.84)$$

where $k' = k_1 k_2 / k_{-1}$ and the reaction is first-order in A. If $k_2 \gg k_{-1}[M]$, the rate becomes

$$-\frac{d[A]}{dt} = k_1[A][M] \tag{4.85}$$

and, when the third body, M, is another molecule of A, the reaction shows a second-order dependence on A. Many of the details of the Marcus theory can be found in the book by Nicholas (1976).

4.5   FREE-RADICAL OR CHAIN MECHANISMS

The basic concepts of free-radical mechanisms were presented in Chapter 1. Reactions following free-radical mechanisms have reactive intermediates containing unpaired electrons, which are produced by homolytic cleavage of covalent bonds. A method of detecting free radicals was published in 1929 and it is based on the fact that metals such as lead react with free radicals. When heated, tetramethyl lead decomposes,

$$(CH_3)_4 Pb \longrightarrow Pb + 4CH_3 \cdot \tag{4.86}$$

A lead mirror is produced in a heated glass tube when tetramethyl lead is passed through it. Also, the lead mirror in a cool portion of the tube can be removed by passing tetramethyl lead through a hot portion of the tube first to produce $CH_3 \cdot$ radicals. In the cool portion of the tube, the reaction is

$$4CH_3 \cdot + Pb \longrightarrow (CH_3)_4 Pb \tag{4.87}$$

However, if the flow system is arranged so that a long tube is used and considerable distance separates the point where the $CH_3 \cdot$ radicals are generated and they react with the cool lead mirror, the reaction is hindered because of radical recombination.

$$2CH_3 \cdot \longrightarrow C_2 H_6 \tag{4.88}$$

Perhaps the best-known example of a chain process, certainly it is the classic case, is the reaction

$$H_2 + Br_2 \longrightarrow 2HBr \tag{4.89}$$

This reaction was studied by Bodenstein and Lind nearly 90 years ago, and the rate law found was written as

$$\frac{d[HBr]}{dt} = \frac{k[H_2][Br_2]^{1/2}}{1 + k'([HBr]/[Br_2])} \tag{4.90}$$

where k and k' are constants with $k' \approx 10$. The [HBr] in the denominator indicates that the rate is decreased as [HBr] increases so that HBr functions as an inhibitor. The reaction has now been studied both thermally and photochemically, and the initiation step is now agreed to be

$$Br_2 \longrightarrow 2Br \cdot \qquad (4.91)$$

The overall reaction scheme was postulated in 1919 by Christiansen, Herzfeld, and Polyani in three separate publications.

The overall process is now described in terms of the elementary steps

$$Br_2 \xrightarrow{k_1} 2Br \cdot \qquad (4.92)$$

$$Br \cdot + H_2 \xrightarrow{k_2} HBr + H \cdot \qquad (4.93)$$

$$H \cdot + Br_2 \xrightarrow{k_3} HBr + Br \cdot \qquad (4.94)$$

$$H \cdot + HBr \xrightarrow{k_4} H_2 + Br \cdot \qquad (4.95)$$

$$2Br \cdot \xrightarrow{k_5} Br_2 \qquad (4.96)$$

Simplification of the mathematical problem is achieved by application of the steady-state hypothesis to those species that occur only in the propagation steps. In this case, it is assumed that [Br·] and [H·] are at some low, essentially constant level. Therefore,

$$\frac{d[Br\cdot]}{dt} = 0 \text{ and } \frac{d[H\cdot]}{dt} = 0$$

Following the same type of treatment used in cases in Chapter 2, we express [Br·] and [H·] in terms of formation and disappearance. Therefore, using simply [H] and [Br] instead of [H·] and [Br·], we can write from the elementary steps above

$$\frac{d[H]}{dt} = k_2[Br][H_2] - k_3[H][Br_2] - k_4[H][HBr] = 0 \qquad (4.97)$$

and

$$\frac{d[Br]}{dt} = 2k_1[Br_2] - k_2[Br][H_2] + k_3[H][Br_2] + k_4[H][HBr] - 2k_5[Br]^2 = 0 \qquad (4.98)$$

Now we can also write for the production of HBr

$$\frac{d[HBr]}{dt} = k_2[Br][H_2] + k_3[H][Br_2] - k_4[H][HBr] \qquad (4.99)$$

If we subtract Eq. (4.97) from Eq. (4.99),

$$\frac{d[HBr]}{dt} - 0 = k_2[Br][H_2] + k_3[H][Br_2] - k_4[H][HBr]$$

$$- \{k_2[Br][H_2] - k_3[H][Br_2] - k_4[H][HBr]\}$$

$$\frac{d[HBr]}{dt} = 2k_3[H][Br_2] \qquad (4.100)$$

Adding Equations (4.97) and (4.98) gives

$$0 = k_2[Br][H_2] - k_3[H][Br_2] - k_4[H][HBr] + 2k_1[Br_2] - k_2[Br][H_2]$$

$$+ k_3[H][Br_2] + k_4[H][HBr] - 2k_5[Br]^2$$

which simplifies to

$$2k_1[Br_2] - 2k_5[Br]^2 = 0 \qquad (4.101)$$

Solving this equation for [Br] yields

$$[Br] = \{(k_1 / k_5)[Br_2]\}^{1/2} \qquad (4.102)$$

Substitution of this value for [Br] in Eq. (4.97) gives

$$k_2[Br][H_2] - k_3[H][Br_2] - k_4[H][HBr] = k_2\{(k_1 / k_5)[Br_2]\}^{1/2}[H_2]$$

$$- k_3[H][Br_2] - k_4[H][HBr] = 0$$

$$k_2\{(k_1 / k_5)[Br_2]\}^{1/2}[H_2] - [H](k_3[Br_2] + k_4[HBr]) = 0 \qquad (4.103)$$

Solving this equation for [H] yields

$$[H] = \frac{k_2\{(k_1 / k_5)[Br_2]\}^{1/2}[H_2]}{k_3[Br_2] + k_4[HBr]} \qquad (4.104)$$

Substituting this result for [H] in Eq. (4.100) we obtain

$$\frac{d[HBr]}{dt} = 2k_3[H][Br_2] = \frac{2k_3k_2\{(k_1/k_5)[Br_2]\}^{1/2}[H_2][Br_2]}{k_3[Br_2]+k_4[HBr]}$$

$$\frac{d[HBr]}{dt} = \frac{2k_2k_3(k_1/k_5)^{1/2}[H_2][Br_2]^{3/2}}{k_3[Br_2]+k_4[HBr]}$$

(4.105)

Dividing numerator and denominator of the right-hand side of this equation by $k_3[Br_2]$ gives

$$\frac{d[HBr]}{dt} = \frac{2k_2(k_1/k_5)^{1/2}[H_2][Br_2]^{1/2}}{1+\dfrac{k_4[HBr]}{k_3[Br_2]}}$$

(4.106)

If we let $k = 2k_2(k_1/k_5)^{1/2}$ and $k' = k_4/k_3$, Eq. (4.106) has the same form as the empirical rate law shown in Eq. (4.90). We need now to examine the relationships of the constants. The bond energies for the molecular species in Steps 3 and 4 are as follows: H—H, 436; Br—Br, 193; and H—Br, 366 kJ/mol. Therefore, both Steps 3 and 4 are exothermic having enthalpies of about −173 and −70 kJ/mol, respectively. Activation energies for forming the activated complexes [H · · · Br · · · Br] and [H · · · H · · · Br] are very low so there will be almost no temperature dependence on the rates of their formation and the temperature effects could well cancel. Therefore, the ratio k_4/k_3 is constant, having a value of about 10.

Other reactions in the chain process could include

$$H_2 \longrightarrow 2H\cdot$$

(4.107)

but the bond energy for H_2 is about 436 kJ/mol while that for Br_2 is 193 kJ/mol. Consequently, any dissociation involving H_2 would be insignificant compared to the dissociation of Br_2. Likewise, the dissociation of HBr to give H· and Br· would be energetically unfavorable as would the reaction

$$Br\cdot + HBr \longrightarrow H\cdot + Br_2$$

(4.108)

Finally, the reactions

$$H\cdot + H\cdot \longrightarrow H_2$$

(4.109)

$$H\cdot + Br\cdot \longrightarrow HBr$$

(4.110)

can be considered as unlikely at best owing to the very low stationary state concentrations of these H· radicals. There are other arguments against these processes as well.

Radicals are generated, consumed, or propagated by a relatively few types of elementary reactions. Radical generation usually involves the homolytic dissociation of some covalent bond.

$$-X-Y- \longrightarrow -X\cdot + -Y\cdot \qquad (4.111)$$

In this case, the process is a high-energy one so it is usually brought about by thermal, photochemical, or electrical means. The consumption of radicals occurs in termination steps that include processes such as

$$Br\cdot + Br\cdot \longrightarrow Br_2 \qquad (4.112)$$

$$\cdot CH_3 + \cdot CH_3 \longrightarrow C_2H_6 \qquad (4.113)$$

The propagation of radicals can involve a transfer of atoms,

$$XY + Z\cdot \longrightarrow XZ + Y\cdot \qquad (4.114)$$

of which the reaction

$$H\cdot + Br_2 \longrightarrow HBr + Br\cdot \qquad (4.115)$$

is a previously seen example. In other cases, a radical may add to another molecule to produce a different radical.

$$XY + Z\cdot \longrightarrow XYZ\cdot \qquad (4.116)$$

An example of this type of process is

$$H\cdot + C_2H_4 \longrightarrow C_2H_5\cdot \qquad (4.117)$$

All of these processes as well as numerous examples of each type are discussed more fully by Nicholas (1976).

A further complication of chain mechanisms is the process known as *branching*. In this case, one radical results in more than a single radical being produced so the number of radicals present is increasing. This results in an autocatalytic reaction that may, as in the reaction of H_2 and O_2, lead to an explosion under certain conditions. The reaction of H_2 and O_2 is a very complicated process that depends on the pressure of the gases, the temperature, and the type of reaction vessel. Some of the reactions involved under certain conditions are believed to be the following, although other steps may also be involved.

$$H_2 + O_2 \xrightarrow{\text{wall}} 2\,OH\cdot \qquad \text{Initiation} \qquad (4.118)$$

$$\cdot OH + H_2 \longrightarrow H_2O + H\cdot \tag{4.119}$$

Propagation

$$H\cdot + O_2 + M \longrightarrow HO_2\cdot + M \tag{4.120}$$

$$H\cdot + O_2 \longrightarrow OH\cdot + \cdot O\cdot \tag{4.121}$$

Branching

$$\cdot O\cdot + H_2 \longrightarrow OH\cdot + H\cdot \tag{4.122}$$

$$H\cdot \xrightarrow{\text{wall}} H_2 \qquad \text{Termination} \tag{4.123}$$

The overall mechanism is very complex and the reader is refered to other sources dealing with gas phase reactions for details (Nicholas, 1976).

4.6 ADSORPTION

A large number of reactions, many of them of great technological importance, involve the reaction of gases on solid surfaces. For example,

$$N_2 + 3\,H_2 \xrightarrow{\text{Fe, oxides}} 2\,NH_3 \tag{4.124}$$

$$4\,NH_3 + 5\,O_2 \xrightarrow{\text{Pt}} 4\,NO + 6\,H_2O \tag{4.125}$$

$$RCH = CH_2 + H_2 \xrightarrow{\text{Ni, Pd, or Pt}} RCH_2CH_3 \tag{4.126}$$

$$CO + H_2O \xrightarrow{Zn_2,\ Cr_2O_3} CO_2 + H_2 \tag{4.127}$$

are but a few such cases. When a solid catalyzes a reaction, the gaseous reactants are attached in some way before the reaction takes place. *Heterogeneous catalysis* is a process in which a solid has gaseous reactants attached that subsequently react. Consequently, it is necessary to begin a discussion of heterogeneous catalysis by describing the process of adsorption in some detail.

In the interior of a solid lattice, each unit (atom, molecule, or ion) is surrounded on all sides. On the surface, the units are not surrounded on one side and, therefore, they can form bonds to other species. While this process may take place by adsorption of molecules or ions from solutions, we are more concerned here with adsorption of gaseous molecules. The sites on the solid where the gases are adsorbed are called *active sites*. The solid material doing the adsorbing is called the *adsorbent*, and the substance adsorbed is called the *adsorbate*.

Interactions between adsorbates and adsorbents cover a wide range of energies. On the one hand, the interactions may be the result of weak van der Waals forces, while on the other, the bonds may represent strong chemical bonding of the adsorbate to the adsorbent. The distinction is not always a clear one, but physical adsorption (*physisorption*) is generally associated with heats of adsorption of 10 to 25 kJ/mol, while chemical adsorption (*chemisorption*) is associated with heats of adsorption of 50 to 100 kJ/mol. In either case, there is presumed to be a relationship between energy and the relative adsorbent/adsorbate position similar to that shown in Figure 4.8.

In general, it is believed that in cases of physical adsorption, the bonding to the surface is so weak that the adsorbent molecules are changed only very slightly

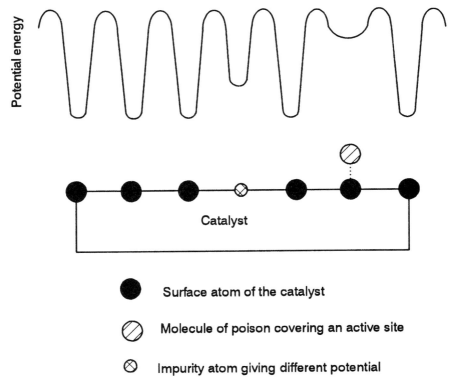

Figure 4.8   Variation in potential energy near a catalyst surface.

by the process. Therefore, physical adsorption does not weaken the bonds in the adsorbate molecules and the adsorbent does not function as a catalyst.

On an atomic scale, adsorption can be considered by quantum mechanical techniques. In this treatment, it is assumed that the forces between the adsorbate and the adsorbent are essentially *chemical* in nature. In that case, the interaction energy is calculated using techniques that are the standard ones in molecular quantum mechanics. However, the energy of a molecule being adsorbed on the surface of a solid is related to distance from the adsorbing site in such a way that the relationship results in a potential energy curve similar to the Morse potential for a diatomic molecule (Figure 4.3). Calculations should produce curves of similar shape, and the calculated energies should match the measured energies. This is a rather formidable task and the results are not always good. Significant progress has been made in this area using extended Huckel molecular orbital (EHMO), self-consistent field (SCF), and complete neglect of differential overlap (CNDO) approaches. A brief review of these results has been given by White (1990). We will now turn our attention to describing the process of adsorption from a bulk macroscopic point of view.

4.6.1 Langmuir Adsorption Isotherm

For chemisorption, one of the most successful approaches for describing the quantitative relationships is that of Irving Langmuir. In this approach, it is assumed that the adsorption process is taking place isothermally and that the uniform adsorbent surface can be covered with a monolayer of adsorbate. Further, it is assumed that there is no interaction between adsorbed molecules and that the available sites all have the same affinity for the gaseous adsorbate.

If the area of the adsorbent is represented as A and the fraction of the surface that is covered by adsorbate is f, we can derive the relationship for adsorption as follows. For an equilibrium of adsorption, we can let the rate of condensation be equal to the rate of evaporation. The rate of evaporation will be proportional to f, the fraction of the surface covered while the rate of condensation will be proportional to $(1 - f)$, the fraction of the surface which is uncovered, and the pressure of the gas. Therefore,

$$k_c (1 - f) P = k_e f \qquad (4.128)$$

If this equation is solved for f, we obtain

$$f = \frac{k_c P}{k_e + k_c P} \qquad (4.129)$$

Dividing both the numerator and denominator of the right-hand side of Eq. (4.129) by k_e and letting $K = k_c/k_e$ gives

$$f = \frac{KP}{1+KP} \tag{4.130}$$

This relationship, known as the Langmuir isotherm, is shown graphically in Figure 4.9. Taking the reciprocal of both sides of Eq. (4.130) gives

$$\frac{1}{f} = \frac{1+KP}{KP} = 1 + \frac{1}{KP} \tag{4.131}$$

Therefore, when $1/f$ is plotted versus $1/P$, a straight line having a slope of $1/K$ and an intercept of 1 results, as is shown in Figure 4.10.

The volume of gas adsorbed is proportional to the fraction of the surface covered,

$$V = V_o f \tag{4.132}$$

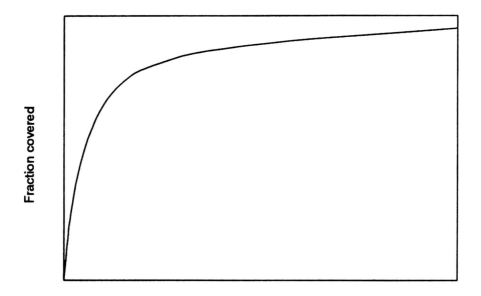

Pressure

Figure 4.9   The Langmuir isotherm for chemisorption.

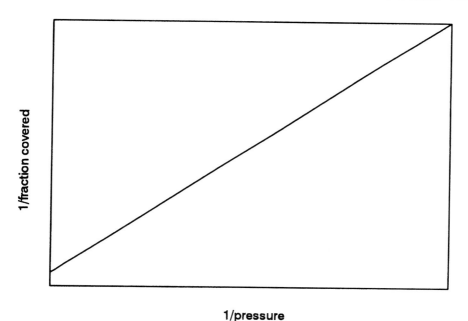

Figure 4.10   Reciprocal plot for Langmuir isotherm.

If the maximum volume adsorbed, V_m, represents complete coverage of the surface,

$$V_m = V_o A \qquad (4.133)$$

and if we let the area be equal to unity (a unit area), then $A = 1$ and $V/V_m = f$. Therefore,

$$\frac{V}{V_m} = f = \frac{KP}{1+KP} \qquad (4.134)$$

which is the Langmuir isotherm. This relationship provides the basis for the volumetric measurement of moles of gas adsorbed as a function of pressure.

If two gases, A and B, are being adsorbed, the fraction of the surface area that is uncovered is $1 - f_A - f_B$. If we describe the rate of condensation of A as

$$\text{Condensation rate} = k_c P_A (1 - f_A - f_B) \qquad (4.135)$$

then the rate of evaporation of A can be expressed as

$$\text{Evaporation rate} = k_e f_A \qquad (4.136)$$

At equilibrium the rates will be equal so we can write

$$k_e f_A = k_c P_A (1 - f_A - f_B) \tag{4.137}$$

or, since $k_c / k_e = K_A$

$$K_A P_A = \frac{f_A}{(1 - f_A - f_B)} \tag{4.138}$$

For gas B, the corresponding equation is

$$K_B P_B = \frac{f_B}{1 - f_A - f_B} \tag{4.139}$$

Therefore, the fraction covered by A and B can be found by solving these equations for f_A and f_B. We will illustrate this procedure by solving for f_A.

Equation (4.138) can be written as

$$f_A = K_A P_A (1 - f_A - f_B) = K_A P_A - f_A K_A P_A - f_B K_A P_A \tag{4.140}$$

Solving this equation for f_B gives

$$f_B = \frac{K_A P_A - f_A K_A P_A - f_A}{K_A P_A} \tag{4.141}$$

Equation (4.139) can be written as

$$f_B = K_B P_B (1 - f_A - f_B) \tag{4.142}$$

Therefore, substitution for f_B gives

$$\frac{K_A P_A - f_A K_A P_A - f_A}{K_A P_A} = K_B P_B \left[1 - f_A - \frac{K_A P_A - f_A K_A P_A - f_A}{K_A P_A} \right]$$

$$\frac{K_A P_A - f_A K_A P_A - f_A}{K_A P_A} = K_B P_B \left[\frac{K_A P_A - f_A K_A P_A - K_A P_A + f_A K_A P_A + f_A}{K_A P_A} \right]$$

Multiplying both sides of this equation by $K_A P_A$ and simplifying gives

$$K_A P_A - f_A K_A P_A - f_A = f_A K_B P_B \tag{4.143}$$

Collecting terms containing f_A and factoring out f_A we obtain

$$f_A (1 + K_A P_A + K_B P_B) = K_A P_A$$

Solving for f_A, we obtain

$$f_A = \frac{K_A P_A}{1 + K_A P_A + K_B P_B} \tag{4.144}$$

Similarly, we find that

$$f_B = \frac{K_B P_B}{1 + K_A P_A + K_B P_B} \tag{4.145}$$

If P_B is 0 or if B interacts weakly with the adsorbent ($K_B \approx 0$)

$$f_A = \frac{K_A P_A}{1 + K_A P_A} \tag{4.146}$$

which is equivalent to Eq. (4.130).

4.6.2   B-E-T Isotherm

The relationship between extent of adsorption and gas pressure shown in Figure 4.9 is by no means the only relationship known. Strictly, it applies only when maximum adsorption results in a monolayer of adsorbate on the surface. Two of the other types of adsorption behavior are shown in Figure 4.11. These processes are associated with the formation of multilayers of adsorbate. The equation that can be derived to describe the formation of multilayers is

$$\frac{P}{V(P^\circ - P)} = \frac{1}{V_m c} + \left(\frac{c-1}{V_m c}\right)\frac{P}{P^\circ} \tag{4.147}$$

where V is the volume of adsorbed gas at standard conditions, P is the pressure of the gas, P° is the saturated vapor pressure of the adsorbate, V_m is the volume of adsorbate at standard conditions required to give a monolayer, and c is a constant. This equation, known as the B-E-T isotherm, is named after Brunauer, Emmett, and Teller, who developed it. The constant c is related to the heat of adsorption of a monolayer, E_{ad}, and the heat of liquefaction of the gas, E_{liq}, by the relationship

$$c = \exp(E_{ad} - E_{liq})/RT \tag{4.148}$$

An adsorption isotherm of the type shown as Curve A in Figure 4.11 results when $E_{ad} > E_{liq}$ and an isotherm of the type shown as Curve B in Figure 4.11 corresponds to the case where $E_{liq} > E_{ad}$. While the B-E-T approach is more successful

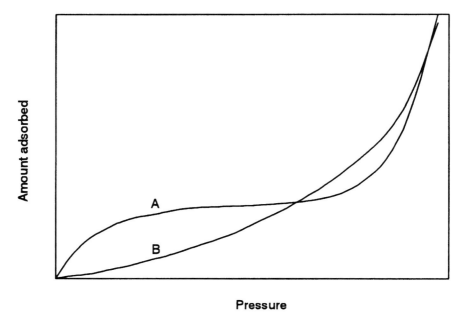

Pressure

Figure 4.11 Adsorption isotherms for formation of multilayers.

in dealing with more complex adsorption cases, we will not discuss its application further. Details on the derivation and use of the equation can be found in the book by White (1990).

4.6.3 Poisons and Inhibitors

For many catalysts, very small amounts of certain substances greatly reduce the effectiveness of the catalysts. These substances are usually designated as *poisons* or *inhibitors*. In some cases, the action of the poison persists only as long as the poison is present in contact with the catalyst. The poison may be one of the products of the reaction, in which case the concentration or pressure of the substance appears in the denominator of the rate law. The poison is adsorbed more strongly than the reactants, but once it is removed the catalyst recovers its activity.

Permanent catalyst poisoning occurs when some material reacts with the catalyst to form a chemically altered surface that no longer retains catalytic properties. A wide range of cases of this type exist. Compounds containing silicon, lead (do not use lead-containing gasoline in an automobile with a catalytic converter), sulfur, arsenic, phosphorus, etc., along with H_2S and CO, are particularly effective poisons toward metallic catalysts. Some of these poisons also inhibit enzyme action and are toxic to animals as well (Chapter 6).

Figure 4.8 shows a poison atom or molecule occupying a site on a solid catalyst. Because of that interaction, there is a very small residual potential for binding an adsorbate. Figure 4.8 is in some ways misleading in that not every surface atom is an active site. The fact that very small amounts of poisons can destroy catalytic activity suggests that the catalytic activity is confined to a rather small fraction of the total surface.

We saw earlier that when a second gas is competing with the reactant for the active sites on the catalyst, the fraction of the catalyst covered by the reactant (A) was decreased. If the inhibitor or poison is designated as X, we find

$$f_A = \frac{K_A P_A}{1 + K_A P_A + K_X P_X} \tag{4.149}$$

If the inhibitor has a large equilibrium constant for adsorption, $1 + K_X P_X \gg K_A P_A$ and

$$f_A = \frac{K_A P_A}{1 + K_X P_X} \tag{4.150}$$

The rate of the reaction of A (R) will be kf_A or

$$R \approx \frac{k K_A P_A}{1 + K_X P_X} \tag{4.151}$$

and at sufficiently high pressures of X,

$$R \approx \frac{k K_A P_A}{K_X P_X} \tag{4.152}$$

While the reaction is first-order in reactant A, the rate law contains the inhibitor function in the denominator showing that the rate is decreased as the amount of inhibitor increases, in accord with our previous statement.

4.7   CATALYSIS

In reactions that are catalyzed by solid surfaces, it is the amount of *adsorbed* gas that determines the rate of the reaction. Therefore, the rate is proportional to the fraction of active sites covered, f.

$$\text{Rate} = kf \tag{4.153}$$

From the Langmuir isotherm (Eq. (4.130)), we determine that

$$\text{Rate} = k\frac{KP}{1 + KP} \tag{4.154}$$

If the reactant gas is one that is strongly adsorbed or if the pressure is high, the fraction of the surface covered approaches unity and KP >> 1 so

$$\text{Rate} = kf \approx k \tag{4.155}$$

Therefore, the rate is independent of the pressure of the reacting gas and the reaction appears to be zero-order.

When the gas is only weakly adsorbed or the pressure is low, 1 >> KP and

$$\text{Rate} = \frac{kKP}{1 + KP} \approx kKP \tag{4.156}$$

which shows the reaction to be first-order in the gaseous reactant. Figure 4.12 shows the behavior of reaction kinetics in these two limiting cases.

In many instances, the progress of a gaseous reaction is followed by the change in pressure of the reacting gas. If the reactant is A and its pressure is P_A, the rate equation is

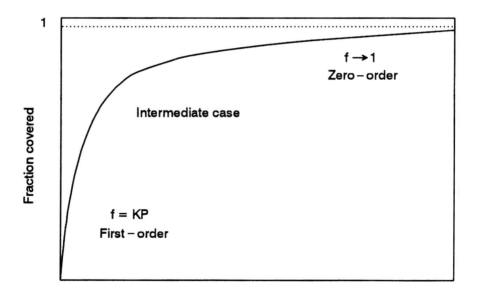

Figure 4.12 Kinetics of surface reactions based on the Langmuir isotherm.

$$-\frac{dP_A}{dt} = k'P_A \tag{4.157}$$

where $k' = kK$. By integration of this equation, we obtain

$$\ln\frac{P_{A,o}}{P_A} = k't \tag{4.158}$$

Such a rate law has been found to correctly model many reactions taking place on solid surfaces.

In the case where the gas is strongly adsorbed or the pressure is high, Eq. (4.157) shows that

$$-\frac{dP_A}{dt} = k \tag{4.159}$$

and

$$P_{A,o} - P_A = kt \tag{4.160}$$

This zero-order rate law has also been found to correctly model certain surface reactions.

Unfortunately, there are cases where neither of the limiting rate laws adequately represents the reaction as is illustrated in Figure 4.12. For such cases,

$$\text{Rate} = kf = \frac{kKP}{1 + KP} \tag{4.161}$$

For a reactant A, the change in pressure of the gas can be used to measure the rate so that

$$-\frac{dP_A}{dt} = \frac{kKP_A}{1 + KP_A} \tag{4.162}$$

which can be written as

$$-\frac{(1 + KP_A)}{KP_A}dP_A = k \tag{4.163}$$

or

$$-\frac{dP_A}{KP_A} - dP_A = k\,dt \tag{4.164}$$

This equation can be simplified to give

$$-\frac{1}{K} \cdot \frac{dP_A}{P_A} - dP_A = k\, dt$$

This equation must be integrated between the limits of $P_{A,o}$ at time equal zero and P_A at time t. Therefore,

$$-(1/K)(\ln P_A - \ln P_{A,o}) - (P_A - P_{A,o}) = kt$$

which can be written as

$$\frac{1}{K} \ln \frac{P_{A,o}}{P_A} + (P_{A,o} - P_A) = kt \qquad \textbf{(4.165)}$$

The similarity of this equation to Eq. (6.53) should be noted because these equations illustrate the similarity between reactions of adsorbed gases and substrates bound to enzyme active sites.

In some cases, kinetic analysis of reactions involving adsorption of gases is carried out using the Freundlich isotherm,

$$f = kP^n \qquad \textbf{(4.166)}$$

where f is the fraction of the surface covered, P is the pressure of the gas, and k and n are constants. Rate laws for intermediate cases can be developed using this approximation.

REFERENCES FOR FURTHER READING

Benson, S. W. (1960) *The Foundations of Chemical Kinetics*, McGraw-Hill, New York, Chapters 7–13, 17. An advanced exposition of gas phase reaction theory.

Bond, G. C. (1987) *Heterogeneous Catalysis: Principles and Applications*, Clarendon Press, Oxford. An excellent introductory book that describes numerous industrial applications of catalysis.

Eyring, H., Eyring, E. M. (1963) *Modern Chemical Kinetics*, Reinhold, New York. A small book that gives a thorough treatment of transition state theory.

Laidler, K. J. (1965) *Chemical Kinetics*, 2d ed., McGraw-Hill, New York, Chapters 4 and 6. A standard coverage of gas phase reaction dynamics.

Marcus, R. A. (1952) *J. Chem. Phys.*, 20, 359.

Maron, S. H., Prutton, C. F. (1965) *Principles of Physical Chemistry*, 4th ed., Macmillan, New York, Chapter 20. Presents a good introduction to adsorption.

Nicholas, J. (1976) *Chemical Kinetics*, Wiley, New York, Chapters 2, 5, 6, and 7.
A clear, thorough coverage of gas phase reaction kinetics.
Truhlar, D. G., Garrett, B. C. (1980) *Acc. Chem. Res.*, 13, 440.
White, M. G. (1990) *Heterogeneous Catalysis*, Prentice Hall, Englewood Cliffs, N. J.,
Chapters 1, 3, 7, 8, and 9. Detailed coverage of the subject.

PROBLEMS

1. Unimolecular decompositions can appear to be first- or second-order under certain conditions. What assumptions were applied to the solution of this problem? Write out the mechanism for the unimolecular decomposition of $X(g)$ and derive the rate law. Explain how this rate law accounts for the observations on reaction order.

2. Suppose that a reaction follows the scheme

$$X \xrightarrow{k_1} R\cdot$$

$$R\cdot + X \xrightarrow{k_2} P + 2\,R\cdot$$

$$R\cdot \xrightarrow{k_3} Z$$

What is a reaction scheme like this called? Derive the rate expression giving $[R\cdot]$ as a function of time. Using your derived rate expression, explain what can happen when the concentration of X is varied between rather wide limits. What type of chemical event does this correspond to?

3. Consider the decomposition of CH_3CHO into CH_4 and CO, which is believed to take place in the steps

$$CH_3CHO \xrightarrow{k_1} CH_3\cdot + CHO$$

$$CH_3\cdot + CH_3CHO \xrightarrow{k_2} CH_4 + CO + CH_3\cdot$$

$$2\,CH_3\cdot \xrightarrow{k_3} C_2H_6$$

Use the steady-state approximation to derive the rate law for the formation of CH_4. What would the rate of formation of CO be?

4. Consider the reaction scheme

$$NO(g) + H_2(g) \underset{k_{-1}}{\overset{k_1}{\rightleftharpoons}} H_2ON(g) \qquad \text{(fast)}$$

$$H_2ON(g) + NO(g) \xrightarrow{k_2} N_2(g) + H_2O_2(g) \qquad \text{(slow)}$$

$$H_2O_2(g) + H_2(g) \xrightarrow{k_3} 2\ H_2O(g) \qquad \text{(fast)}$$

Write the rate law for the overall reaction in terms of the steps given above. Apply the steady-state approximation and obtain the final rate law.

5. The reaction between H_2 and Br_2 has been described in terms of these steps.

$$Br_2 \underset{k_{-1}}{\overset{k_1}{\rightleftharpoons}} 2\ Br \cdot$$

$$Br \cdot + H_2 \xrightarrow{k_2} HBr + H \cdot$$

$$H \cdot + Br_2 \xrightarrow{k_3} HBr + Br \cdot$$

$$H \cdot + HBr \xrightarrow{k_4} H_2 + Br \cdot$$

Write the rate laws for the change in concentration of $H\cdot$, $Br\cdot$, and HBr with time. Apply the steady-state approximation and show that the rate of formation of HBr is

$$\frac{d[HBr]}{dt} = \left(\frac{k_1}{k_{-1}}\right)^{1/2} \frac{2k_2[H_2][Br_2]^{1/2}}{1 + \dfrac{k_4}{k_3}\dfrac{[HBr]}{[Br_2]}}$$

6. For each of the following, use the Langmuir isotherm and provide an interpretation for the observation. (a) The decomposition of gas X on solid S is zero-order. (b) The decomposition of gas Y on solid S is first-order. (c) The decomposition of NH_3 on a platinum surface is inversely proportional to the pressure of H_2 and directly proportional to the pressure of NH_3.

7. The reaction of NO(g) with $Br_2(g)$ produces ONBr(g) and may take place in the steps

$$NO(g) + Br_2(g) \underset{k_{-1}}{\overset{k_1}{\rightleftharpoons}} ONBr_2(g) \qquad \text{(fast)}$$

$$ONBr_2(g) + NO(g) \overset{k_2}{\longrightarrow} 2\,ONBr(g) \qquad \text{(slow)}$$

Assuming that $ONBr_2(g)$ establishes a steady-state concentration, derive the rate law for the production of ONBr(g).

8. Suppose a gaseous reaction takes place in the steps

$$A \underset{k_{-1}}{\overset{k_1}{\rightleftharpoons}} B$$

$$B + C \overset{k_2}{\longrightarrow} D$$

Derive the rate law for the formation of D and show the limiting forms at high pressure and low pressure.

9. The decomposition of tetraborane-10, B_4H_{10}, is thought to take place in the following steps (Bond, A. C., Pinsky, M. L., *J. Am. Chem. Soc.* 1970, 92, 32).

$$B_4H_{10} \overset{k_1}{\longrightarrow} B_3H_7 + BH_3$$

$$B_3H_7 \overset{k_2}{\longrightarrow} BH_2 + B_2H_5$$

$$B_2H_5 \overset{k_3}{\longrightarrow} BH_2 + BH_3$$

$$BH_2 + B_4H_{10} \overset{k_4}{\longrightarrow} B_4H_9 + BH_3$$

$$B_4H_9 \overset{k_5}{\longrightarrow} B_2H_5 + B_2H_4$$

$$B_2H_4 \overset{k_6}{\longrightarrow} H_2 + \text{solid product}$$

$$2 \quad BH_2 \xrightarrow{k_7} B_2H_4$$

$$2 \quad BH_3 \xrightarrow{k_8} B_2H_6$$

Make the steady-state assumption regarding BH_2, BH_3, B_2H_4, B_2H_5, B_3H_7, and B_4H_9, and show that the rate law is

$$-\frac{d[B_4H_{10}]}{dt} = k_1[B_4H_{10}] + \frac{k_4 k_1^{1/2}}{k_7^{1/2}}[B_4H_{10}]^{3/2}$$

chapter 5

REACTIONS IN SOLUTIONS

There are many gaseous materials which react, and there are a large number of reactions which take place in the solid state. In spite of this, most chemical reactions are carried out in solutions, with a large fraction of chemistry taking place in aqueous solutions. However, before one undertakes a description of the effects of the solvent on a reaction, it is necessary to describe some of the characteristics of liquids and solutions.

5.1 THE NATURE OF LIQUIDS

It is a relatively simple process to model the behavior of most gases satisfactorily (except at high pressures or low temperatures) using kinetic theory. For many gases, the interactions between molecules can even be ignored. The interactions between ions in ionic solids are adequately treated using Coulomb's law because the interactions are electrostatic in nature. While molecular motions in gases are random, solids have units (ions, atoms, or molecules) that are localized to fixed positions except for vibrations. Part of the problem in dealing with the liquid state is that there are intermolecular forces that are too strong to ignore, but that are not strong enough to restrict molecular motion completely. There is local structure within clusters of molecules, but there is rather extensive interchange between clusters. This view has been called the *significant structure theory* of liquids.

While the problem of structure and order in the liquid state is considerable, another problem is that of which force law to use to describe the intermolecular interactions. Overall, the molecules are neutral, but there may be charge separations within the molecules. Therefore, dipole-dipole forces may be the dominant type of interaction between molecules. On the other hand, molecules such as CCl_4 are nonpolar so there must be other types of forces responsible for the properties of the liquid. Because the interactions between molecules of liquids provide a basis for discussing the nature of solutions, we will begin this chapter with a brief discussion of intermolecular forces.

5.1.1   Intermolecular Forces

If a diatomic molecule is composed of two atoms having different electronegativities, the molecule will be polar. The shared electrons will spend a greater fraction of time around the atom having the higher electronegativity. A measure of this charge separation is μ, the dipole moment,

$$\mu = q \cdot r \tag{5.1}$$

where q is the quantity of charge separated and r is the distance of separation. The quantity of charge separated will be a fraction of the electron charge (4.8×10^{-10} esu or 1.6×10^{-19} C), and the distance of separation will be on the order of 10^{-8} cm. Therefore, $q \cdot r$ will be on the order of 10^{-18} esu $\cdot$ cm, and it is convenient to measure μ in units of this size: 10^{-18} esu $\cdot$ cm = 1 Debye (abbreviated as D).

In 1912, Keesom considered polar molecules to be assemblies of charges although there is no *net* charge. The assembly of + and – charges generates an electric field that depends on the distribution of charge within the molecule. The potential energy of the interaction of the dipoles depends on the orientation of the dipoles. For two polar molecules having dipole moments μ_1 and μ_2, the interaction energy is given by

$$E_D = -\frac{\mu_1 \mu_2}{r^3} (2\cos\theta_1 \cos\theta_2 - \sin\phi_1 \sin\phi_2 \cos(\phi_1 - \phi_2)) \tag{5.2}$$

where θ_1, θ_2, ϕ_1, and ϕ_2 are the angular coordinates (in polar coordinates) giving the orientations of the two dipoles, and r is the average distance of separation. The extremes of interaction of two dipoles can be shown as

Repulsion   Attraction

These two extremes give rise to factors of +2 and –2 for repulsion and attraction, respectively. However, there is an effect of thermal energy that prevents the total population of the lower energy (attraction) state. At sufficiently high temperatures, the attraction is completely overcome and the orientation of the dipoles is random. If all possible orientations between these two extremes are considered, no net attraction results. At intermediate temperatures, there is a greater population of the orientation leading to attraction, which results in some average preferred orientation and a net attraction results. The orientation energy, ΔE, involves a Boltzmann population of two states of different energy, and it is, therefore, temperature dependent. It involves a factor of $e^{-E/kT}$ where **k** is Boltzmann's constant. It can be shown that this energy varies as μ^2/r^3, and assuming that $\Delta E < kT$, the equation that results can be written for two molecules having dipole moments μ_1 and μ_2 as

$$E_D = -\frac{2\mu_1{}^2\mu_2{}^2}{3r^6kT} \qquad (5.3)$$

If the two dipoles are identical, this equation becomes

$$E_D = -\frac{2\mu^4}{3r^6kT} \qquad (5.4)$$

If the energy per mole is to be considered, **k** is replaced by R, since **k** is R/N where N is Avogadro's number.

In solutions containing polar molecules, the solvent strongly affects the association of dipoles. In general, if the solvent has low polarity and/or dielectric constant, the dipoles will be more strongly associated. If the solvent is also polar, it is likely that the solvation of each polar solute molecule will be strong enough that solute molecules will be unable to interact with each other. Thus, the association of a polar solute, D,

$$2D \rightleftharpoons D_2 \qquad (5.5)$$

for which the equilibrium constant is $K = [D_2]/[D]^2$, will be strongly solvent dependent. In the same way that interactions between ions are governed by Coulomb's law, the solvent will affect the attraction between dipoles. Species that are of extreme difference in polarity may not be completely miscible owing to each type of molecule interacting strongly with its own kind. Although they are weak compared to chemical bonds, dipole-dipole forces are of considerable importance.

A permanent dipole, $\mu = q \cdot r$, induces a charge separation in a neighboring molecule that is proportional to the polarizability of the molecule. If the polarizability of the molecule is represented as α, the energy of the interaction can be expressed as

$$E_I = -\frac{2\alpha\mu^2}{r^6} \qquad (5.6)$$

These forces between polar molecules and those having a dipole induced in them are called *dipole-induced dipole forces* and they are essentially temperature independent.

It is immediately obvious that there must be *some* force between molecules that are nonpolar because CH_4, He, CO_2, and similar molecules can be liquified. These forces must be electrical in nature, but can not be the result of an overall charge separation within the molecule. If we consider two helium atoms as shown in Figure 5.1, it is possible that at some instant both of the electrons in one atom will be on the same side of the atom. There is an *instantaneous* dipole that will cause an instantaneous change in the electron distribution in the neighboring atom. There will exist, then, a weak force of attraction between the two atoms. Such forces between instantaneous dipoles are the *London* (or *dispersion*) *forces.* They can be thought of as weak forces between the nuclei in one molecule and the electrons in another.

The energy of interaction by this means can be written as

$$E_L = -\frac{3h\upsilon_o\alpha^2}{4r^6} \tag{5.7}$$

where υ_o is the frequency of the zero-point vibration and α is the polarizability. Because $h\upsilon_o$ is approximately equal to the ionization energy of the molecule, I, Eq. (5.7) becomes

$$E_L = -\frac{3I\alpha^2}{4r^6} \tag{5.8}$$

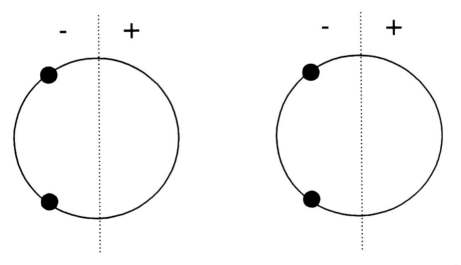

Figure 5.1 Two helium atoms showing instantaneous dipoles which result in a weak force of attraction.

If two different types of molecules are involved, the interaction energy is

$$E_L = -\frac{3}{2} \frac{h\alpha_1\alpha_2 \upsilon_1 \upsilon_2}{r^6(\upsilon_1 + \upsilon_2)} \tag{5.9}$$

When written in terms of the ionization potentials, this equation can be written as

$$E_L = -\frac{3I_1 I_2 \alpha_1 \alpha_2}{2r^6(I_1 + I_2)} \tag{5.10}$$

While it may be somewhat surprising, many molecules of greatly differing structures have ionization potentials which are approximately constant. The examples shown in Table 5.1 illustrate this. Therefore, the product of I_1 and I_2 is sometimes replaced by a constant.

Slater and Kirkwood have derived an expression for the London energy that makes use of the number of outermost electrons in the molecule. That expression is

$$E_L = -\frac{3he\alpha^2}{8\pi r^6 m^{1/2}} \left(\frac{n}{\alpha}\right)^{1/2} \tag{5.11}$$

where e is the charge on the electron, and n is the number of electrons in the outermost shell.

Since the London attraction energy depends on the magnitude of α, it shows a general relationship to molecular size. For example, the boiling point of a liquid involves the separation of molecules from their nearest neighbors. Thus, the boiling points of a given series of compounds (e.g., the hydrocarbons, C_nH_{2n+2}) show a general increase in boiling point as n (or α) increases. Similarly, the halogens reflect this trend with F_2 and Cl_2 being gases at room temperature while Br_2 is a liquid, and I_2 is a solid. All are nonpolar, but the number of electrons increases for the series and the polarizability depends on the ability to distort the electron cloud of the molecule. Generally, the polarizability of molecules increases as the number of electrons increases, but it is important to note that molecules which have delocalized electron density have mobile electrons

Table 5.1 Ionization potentials for several molecules.

Molecule	I.P. (ev)	Molecule	I.P. (ev)
Acetone	9.69	Methanol	10.85
Benzene	9.24	3-Methylpentane	10.06
n-Butane	10.63	Pyrazine	10.00
1,4-Dioxane	9.13	Sulfur dioxide	11.7

and generally have mobile electron clouds. Such electron clouds can be distorted giving rise to high polarizability.

It is also important to note that London forces play an important role in contributing to the overall stability of crystal lattices. Even though the dominant force is the Coulombic force between oppositely charged ions, the London forces are significant in the case of large, polarizable (*soft* in terms of the hard-soft interaction principle) ions. For example, in AgI the Coulomb attraction is 808 kJ mol^{-1} and the London attraction amounts to 128.7 kJ mol^{-1}. As expected, these forces are much less important for crystals like NaF because the polarizabilities of the ions are so small. Because of the variation with $1/r^6$, the London attraction is restricted to very short distances (nearest neighbors).

Various equations have been used to represent the repulsion that exists between molecules at short distances. One such equation is

$$E_r = ae^{-br} \qquad (5.12)$$

where a and b are constants. One type of potential function that results is the Mie potential,

$$E = \frac{A}{r^n} - \frac{B}{r^m} \qquad (5.13)$$

where A, B, *m*, and *n* are constants. The repulsion (positive) term is written as

$$E_r = \frac{j}{r^n} \qquad (5.14)$$

where *n* is in the range of 9 to 12. The Lennard-Jones potential combines this form with $1/r^6$ for the attraction and usually uses $n = 12$ as the exponent in the repulsion term. The resulting equation is

$$E = \frac{j}{r^{12}} - \frac{k}{r^6} \qquad (5.15)$$

which is referred to as the "6–12" or Lennard-Jones potential.

5.1.2   The Solubility Parameter

It is readily apparent that a liquid has cohesion, which holds the liquid together. It should be apparent that the energy with which the liquid is held together is related to the heat necessary to vaporize it. In fact, the cohesion energy, E_c, is given by

$$E_c = \Delta H_v - RT \qquad (5.16)$$

The work done as the vapor expands against the external (atmospheric) pressure is RT. The quantity E_c/V is called the cohesion energy *density* because it is the cohesion energy per unit volume. The thermodynamic relationship

$$dE = TdS - PdV \qquad (5.17)$$

provides a way of interpreting the cohesion energy. From Eq. (5.17), we obtain

$$\frac{\partial E}{\partial V} = T\left(\frac{\partial S}{\partial V}\right)_T - P = T\left(\frac{\partial P}{\partial T}\right)_V - P \qquad (5.18)$$

where P is the *external pressure*. The *internal pressure* is given by

$$P_i = T\left(\frac{\partial P}{\partial T}\right)_V = T\frac{\alpha}{\beta} \qquad (5.19)$$

where α is the coefficient of thermal expansion and β is the compressibility. We can also write

$$\left(\frac{\partial P}{\partial T}\right)_V = \frac{(\partial V/\partial T)_P}{(\partial V/\partial P)_T} \qquad (5.20)$$

The numerator of the right-hand side is the coefficient of thermal expansion, α, and the denominator is the coefficient of compressibility, β. Therefore,

$$P_i = T\frac{\alpha}{\beta} \qquad (5.21)$$

For most liquids, the internal pressure ranges from 2000 to 8000 atm. Therefore,

$$E_c = P_i - P \approx P_i \qquad (5.22)$$

because $P_i \gg P$.

The cohesion energy (*energy* of vaporization) per unit volume is obtained from E_c/V where V is the molar volume. For example, if two liquids have the same value of E_c/V, the heat of mixing is zero (ideal solution). If the E_c/V values are not equal, the heat of mixing will be positive (nonideal solution). However, in developing the theory of solutions, the quantity $(E_c/V)^{1/2}$ is encountered. This quantity is known as the *solubility parameter*, δ. The solubility parameter is given in $(\text{cal}/\text{cm}^3)^{1/2}$ or in $(\text{J}/\text{cm}^3)^{1/2}$. The unit of 1 $\text{cal}^{1/2}/\text{cm}^{3/2}$ is called 1 *Hildebrand* (*h*) in honor of Joel Hildebrand, who did extensive work on the nature of solutions. Table 5.2 shows solubility parameters for several common solvents. It can be seen that values for δ range from about 7 $(\text{cal}/\text{cm}^3)^{1/2}$ for hexane to about four times this value for a strongly associated liquid such as water.

Table 5.2   Solubility parameters for selected liquids.

Liquid	Solubility parameter (h)	Liquid	Solubility parameter (h)
C_6H_{14}	7.3	CS_2	10.0
CCl_4	8.6	CH_3NO_2	12.6
C_6H_6	9.1	Br_2	11.5
$CHCl_3$	9.3	$HCON(CH_3)_2$	12.1
$(CH_3)_2CO$	9.76	C_2H_5OH	12.7
$C_6H_5NO_2$	11.6	H_2O	26.0

Since the cohesion energy is given by

$$E_c = \Delta H_v - RT \tag{5.23}$$

we can calculate it when ΔH_v is known. Values for the heat of vaporization of many liquids are tabulated in handbooks. However, the heat of vaporization can be obtained from

$$\frac{d \ln P}{d(1/T)} = \Delta H_v \tag{5.24}$$

if we have the vapor pressure expressed as a function of temperature. While there are many equations used to relate vapor pressure to temperature, one of the most convenient for this use is the Antoine equation,

$$\log P = A - \frac{B}{C+t} \tag{5.25}$$

where A, B, and C are constants characteristic of the liquid, and t is the temperature in °C. Antoine constants have been tabulated for a large number of liquids. If the vapor pressure is known at several temperatures, the Antoine constants can be calculated. Using the Antoine equation and Eq. (5.24), we find that

$$E_c = RT\left(\frac{2.303BT}{(C+t)^2} - 1\right) \tag{5.26}$$

Therefore, the solubility parameter can be calculated if the molar volume is known. The molar volume requires the density at the desired temperature, and the required density data are usually available for most liquids. If the density, ρ, at the desired temperature is unavailable but it is available at other temperatures, the data can be fitted to the equation

$$\rho = a + bT + cT^2 \tag{5.27}$$

and the constants can be evaluated using a least squares method. The calculated density at the desired temperature and the molar volume can be determined.

Because the solubility parameter reflects the intermolecular forces in a liquid, it is a very useful parameter. Also, the sum of the intermolecular interactions is expected to be the result of dipole-dipole forces, London dispersion forces, and hydrogen bonding. Therefore, it is possible to separate the contributions so that

$$\delta^2 = \delta_L^{\,2} + \delta_D^{\,2} + \delta_H^{\,2} \tag{5.28}$$

where δ_D, δ_L, and δ_H are contributions to the solubility parameter from dipole-dipole, London, and hydrogen-bonding interactions, respectively.

5.1.3 Solvation of Ions and Molecules

When an ionic compound dissolves in a polar solvent such as water, the ions become strongly solvated. The ion-dipole forces produce a layer of solvent molecules (the primary solvation sphere) surrounding each ion. This layer can cause other molecules in the immediate vicinity to become oriented as well. Although the primary solvation sphere may seem to be rather firmly attached to the ion, there is considerable interchange between these molecules and those of the bulk solvent in a dynamic process. For ions like $[Cr(H_2O)_6]^{3+}$, the exchange of coordinated water molecules and those of the bulk solvent is very slow. As we shall see, some desolvation of ionic species may be required before a reaction can take place.

Although the symbol for the solvated proton in acidic solutions is written as H_3O^+, the H^+ is solvated by more than one water molecule. In fact, the dominant species is probably $H_9O_4^+$ or $(H_2O)_4H^+$, and this ion has been identified in vapors above concentrated acids as well as a cation in solids. Other species (e.g., $H_7O_3^+$) also exist in equilibrium with $H_9O_4^+$.

Theoretical treatment of the solvation of ions is quite difficult. If we could use a simple electrostatic approach in which polar molecules interact with a charged ion, the problem would be much simpler. However, the fact that the polar solvent molecules interact with an ion causes their character to change somewhat, and the polarity of the molecules is increased due to the charge separation resulting from ion-dipole forces. Solvent molecules which are bound to an ion have different dipole moment and dielectric constant from the bulk of the solvent. The bound solvent molecules are essentially restricted in their ability to respond to an applied electric field. Therefore, the dielectric constant of the attached water is smaller than that of the bulk solvent.

When an electrostatic approach is used and the dielectric constant, ε, is assumed to be the same as the bulk solvent, the free energy of hydration of an ion of radius r is

$$\Delta G_h = -\frac{NZ^2e^2}{2r}\left(1-\frac{1}{\varepsilon}\right) \tag{5.29}$$

where e is the electronic charge, Z is the charge on the ion, and N is Avogadro's number. However, agreement between calculated and experimental values is usually poor. One way around this is to use an "effective" ionic radius, which is the ionic radius of the ion plus the radius of a water molecule (about 0.75 Å). Another way to improve the calculation is to correct for the altered dielectric constant. When this approach is used, the dielectric constant is expressed as a function of ionic radius. This is done because smaller, more highly charged ions are more strongly solvated and restrict the motion of water molecules to a greater extent. The effective dielectric constant of a liquid changes around an ion in solution, and the higher the charge on the ion, the greater the change. This effect occurs because the dielectric constant is a measure of the ability of a molecule to orient itself in aligning with an electric field. Because the solvent molecules become strongly attached to an ion, they have a reduced ability to orient themselves in the electric field so the dielectric constant is smaller than that for the bulk solvent. Moreover, the reduction in dielectric constant is greater the closer the solvent molecules get to the ion and the higher the charge on the ion.

5.1.4   The Hard-Soft Interaction Principle

We have already alluded to one of the most useful and pervasive principles in all of chemistry, that being the hard-soft interaction principle. This principle relates to many areas, but it is most directly applicable to electron pair donation and acceptance (Lewis acid-base interactions). The terms *hard* and *soft* relate essentially to the polarizability of the species. For example, I^- has a large size so its electron cloud is much more distortable than that of F^-. Likewise, Hg^{2+} is a large metal ion having a low charge, while Be^{2+} is a very small ion. The result is that Hg^{2+} is considered to be a soft Lewis acid while Be^{2+} is considered to be a hard Lewis acid. As a result of these characteristics, Hg^{2+} interacts preferentially with I^- while Be^{2+} interacts preferentially with F^-. The hard-soft interaction principle indicates that *species of similar electronic character (hard or soft) interact best.* It does *not* say that hard Lewis acids will not interact with soft Lewis bases, but the interaction is more favorable when the acid and base are similar in hard-soft character.

The applications of the hard-soft interaction principle are numerous. For example, if we consider the potential interaction of H^+ with H_2O or I^-,

we find that H^+ being a hard electron pair acceptor interacts preferentially with a pair of electrons in an orbital on the oxygen (smaller) atom. Accordingly, HI is completely ionized in dilute aqueous solutions. If we consider the competition for H^+ between F^- and H_2O,

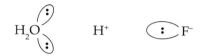

we find that the pairs of electrons on F^- and those of the oxygen atom in a water molecule are contained in orbitals of similar size. Furthermore, the negative charge on the F^- increases the attraction between H^+ and F^-. As a result, H^+ interacts more strongly with F^- than it does with H_2O and, therefore, HF ionizes only slightly and behaves as a weak acid. Further, if we consider the complex formed between Pt^{2+} (low charge, large size, soft electron acceptor) and SCN^-, it is found that the bonding is Pt^{2+}—SCN. The complex of Cr^{3+} (small size, high charge, hard electron acceptor) with SCN^- has the bonding arrangement Cr^{3+}—NCS. These results arise because the sulfur end of SCN^- is considered to be a soft electron donor while the nitrogen end is considered to be a hard electron donor.

The primary reason for discussing the hard-soft interaction principle (HSIP) at this time is because of its usefulness in dealing with solubility and solvation. Certainly, the principle "like dissolves like" has been known for a very long time. We will mention here only a few aspects of the HSIP and its relationship to solubility. As an example, we have already mentioned that NaCl is essentially insoluble in nitrobenzene ($\mu = 4.27$ D). Even though nitrobenzene is quite polar, it can not solvate ions such as Na^+ or Cl^- because of its molecular size. The solubility of NaCl in water and alcohols also shows an interesting trend and allows us to see the effects of solvent properties. The data are shown in Table 5.3.

As the size of the solvent molecules increases and the dielectric constant decreases, the solubility of NaCl decreases. The size and character of the alkyl group becomes dominant over that of the polar —OH group. Accordingly, the solubility of ionic solids decreases with increasing size of the alkyl group.

Table 5.3   Solubility of NaCl in several solvents.

Solvent	H_2O	CH_3OH	C_2H_5OH	$n-C_3H_7OH$
Solubility, mole percent	10.0	0.772	0.115	0.00446
Dipole moment, D	1.84	1.66	1.66	1.68
Dielectric constant, ε	78.5	24.6	20.1	18.3

It has long been known (and utilized) that liquid SO_2 will dissolve aromatic hydrocarbons. The resonance structures of SO_2,

show its delocalized electron density due to π–bonding. Therefore, its similar electronic character to that of aromatic hydrocarbons results in the solubility of these compounds in liquid SO_2. While we have barely introduced the applications of HSIP (often referred to as HSAB when acid-base chemistry is the focus), the suggested readings at the end of this chapter can be consulted for additional details. A great deal of what will be discussed later about the solvation of reactants and transition states can be reduced to applications of this very important and versatile principle.

5.2   SOLVENT POLARITY EFFECTS ON RATES

We have already illustrated the effects of dipole-dipole association. For example, the more strongly solvated an ion or molecule is, the more difficult it is for desolvation to occur so that an active site is exposed. Reactions in which ions are *produced* in forming the transition state will usually be *accelerated as the solvent dielectric constant and dipole moment increase for a series of solvents*. The increased solvation of the ions produced will cause this effect. In contrast, *reactions that involve the combination of ions to produce a transition state of low charge will be retarded by solvents that strongly solvate ions*.

It is generally true that the formation of the transition state involves some change in the distribution of charge in the reactants. Neutral molecules frequently have charge separations induced (see Section 1.5.3). Conversely, forming a transition state during the reaction of ions generally involves cancellation or rearrangement of a portion of the charges.

An early attempt to explain these factors was put forth by Sir Christopher Ingold and his coworkers (1935). The cases considered involve *charge neutralization* as positive and negative ions react and *charge dispersion* as a positive or negative ion has part or all of its charge spread over the transition state. In these cases, the rate of the reaction *decreases* in a series of solvents of *increasing* polarity. On the other hand, a reaction in which a molecule having a symmetric charge distribution forms a transition state having some charge separation will have a rate that increases with solvent polarity. It must be remembered, however, that dipole moment alone is not a good predictor of solvent behavior toward ions. For example, nitrobenzene is quite polar but it is a very poor solvent for materials containing small ions (e.g., NaCl) because of the size of the nitrobenzene molecules. The dipole moment is the product of charge separated and the distance of separation, $q \cdot r$, so a rather large μ could be the result of a small amount of charge separated by a rather large distance. Molecules having these characteristics would not be good solvents for ionic salts containing small ions.

Some solvents consisting of polar molecules solvate anions and cations to different degrees because of their molecular structure. For example, DMF (N, N-dimethylformamide) is polar,

$$\begin{array}{c} H \\ \diagdown \\ \overset{\delta+}{C} = \overset{\delta-}{\bar{O}} \\ \diagup \\ (CH_3)_2 N \end{array}$$

but the positive end of the dipole is not as accessible for solvating anions as is the negative end for solvating cations. This is also true for solvents such as $(CH_3)_2SO$ and CH_3CN. All are good solvents for polar or ionic compounds. Because the negative end of the dipole is less shielded than is the positive end, cations tend to be more strongly solvated than are anions. This results in the anions being able to undergo nucleophilic substitution reactions in those solvents, and the rates tend to be higher than when a solvent such as CH_3OH is used because it solvates both cations and anions about equally well.

The reaction

$$CH_3I + Cl^- \longrightarrow CH_3Cl + I^- \tag{5.30}$$

shows this solvent dependence, and relative rates of the reaction in several solvents are shown in Table 5.4. The number of cases where similar observations on solvent effects are encountered is enormous.

Table 5.4 Relative rates of the reaction shown in Eq. (5.30) in several solvents (Ege, 1994).

Solvent	Relative rate
CH_3OH	1
$HCONH_2$	125
$HCON(CH_3)_2$	1.2×10^6
$(CH_3)CON(CH_3)_2$	7.4×10^6

5.3 IDEAL SOLUTIONS

The behavior of real solutions, such as those in which most reactions take place, is based on a description of *ideal* solutions. The model of an ideal solution is based on Raoult's law. While we can measure the concentration of a species by its mole fraction, X_i, the fact that the solution is not ideal tells us that thermodynamic behavior must be based on the fugacity, f_i. In this development, we will use f_i for the fugacity of the *pure* component i and $\mathbf{f}_i$ as the fugacity of component i in the *solution*. When X_i approaches unity, its fugacity is given by

$$\mathbf{f}_i = X_i f_i \tag{5.31}$$

This is expressed by the relationship known as the Lewis and Randall rule,

$$\lim_{X_i \to 1} (\mathbf{f}_i / X_i) = f_i \tag{5.32}$$

When X_i approaches zero, the limit of $\mathbf{f}_i/X_i$ approaches a constant, C, which is known as Henry's constant.

$$\lim_{X_i \to 0} (\mathbf{f}_i / X_i) = C_i \tag{5.33}$$

When a solution behaves *ideally*, $\mathbf{f}_i = C_i$ for all values of X_i. This means that we can write

$$\mathbf{f}_i = X_i f_i^\circ \tag{5.34}$$

where f_i° is the fugacity of the standard state of component i. Usually, f_i° is taken as the fugacity of the *pure* component i at the temperature and pressure of the *solution*.

When two components are mixed and a solution is formed, the properties of the mixture (solution) are related to those of the individual components and the composition of the solution. For example, the change in volume is described as

$$\Delta V = V_{actual} - \Sigma X_i V_i^\circ \tag{5.35}$$

where V_i° is the molar volume of pure component i in its standard state. If we represent some property, **P**, in terms of the *molar* properties of the components, P_i, we obtain

$$\mathbf{P} = \Sigma X_i P_i \tag{5.36}$$

Therefore, the *change* in the property upon mixing the components can be given by the equation

$$\Delta \mathbf{P}_i = \Sigma X_i (P_i - P_i^\circ) \tag{5.37}$$

where P_i° is the property of the standard state. When the property considered is the free energy, the equation becomes

$$\Delta G = \Sigma X_i (G_i - G_i^\circ) \tag{5.38}$$

Using the relationship that

$$\Delta G = RT \ln (f_i / f_i^\circ) \tag{5.39}$$

mixing causes a change in free energy that is given by

$$\Delta G = RT \Sigma X_i \ln (f_i / f_i^\circ) \tag{5.40}$$

The ratio (f_i/f_i°) is the *activity* of component i in the solution. For a pure component $(X_i = 1)$, the activity of the component in its standard state gives

$$a_i = X f_i^\circ / f_i^\circ \tag{5.41}$$

In an *ideal* solution, since the fugacity of component i is $f_i = X_i f_i^\circ$, we can write

$$a_i = f_i / f_i^\circ = X_i f_i^\circ / f_i^\circ = X_i \tag{5.42}$$

which shows that the activity of component i can be approximated as the mole fraction of i, X_i. Therefore, Eq. (5.40) becomes

$$\Delta G = RT \Sigma X_i \ln X_i \tag{5.43}$$

or

$$\frac{\Delta G}{RT} = \Sigma X_i \ln X_i \tag{5.44}$$

In a similar way, it can be shown that because the composition of an ideal solution is independent of temperature,

$$\frac{\Delta H}{RT} = -\Sigma X_i \left(\frac{\partial \ln X_i}{\partial T} \right)_{P,X} = 0 \tag{5.45}$$

The X subscript after the partial derivative is for $X \ne X_i$. This equation indicates that the heat of solution for an *ideal* solution is zero. The entropy of solution can be shown to be

$$\frac{\Delta S}{R} = -\Sigma X_i \ln X_i \tag{5.46}$$

It is apparent that one of the criteria for the mixture being ideal is that $\Delta H_{mixing} = 0$. However, ΔG_{mixing} and ΔS_{mixing} are not zero. The relationships of the thermodynamic quantities to composition for an ideal solution are shown in Figure 5.2. Real solutions are described in terms of the *difference* between the experimental value for a property and that which would result for an ideal

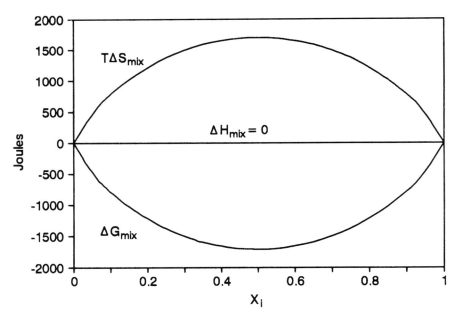

Figure 5.2 Relationship between the composition of an ideal solution and the thermodynamic quantities.

solution at the same conditions. These differences are referred to as the *excess property*,

$$P^E = P_{measured} - P_{ideal} \qquad (5.47)$$

or

$$\Delta P^E = \Delta P_{measured} - \Delta P_{ideal} \qquad (5.48)$$

5.4 COHESION ENERGIES OF IDEAL SOLUTIONS

If the forces between molecules are of the van der Waals type, it can be shown that the internal pressure, P_i, is given by the change in energy with volume at constant temperature,

$$\left(\frac{\partial E}{\partial V} \right)_T = P_i = \frac{a}{V^2} \qquad (5.49)$$

where a/V^2 is the same as is given in the van der Waals equation,

$$\left(P + \frac{n^2 a}{V^2} \right)(V - nb) = nRT \qquad (5.50)$$

While the terms *cohesion energy density* and *internal pressure* refer to the same characteristic of a liquid, they are not identical. The cohesion energy density (E_c/V) is equivalent to the *energy* of vaporization per mole of the liquid and it is calculated in that way. The internal pressure, P_i, is given by

$$P_i = \left(\frac{\partial E}{\partial V} \right)_T = T \left(\frac{\partial S}{\partial V} \right)_T - P \qquad (5.51)$$

Since

$$\left(\frac{\partial S}{\partial V} \right)_T = \left(\frac{\partial P}{\partial T} \right)_V \qquad (5.52)$$

the internal pressure is determined from such thermal pressure coefficient measurements. While P_i and E_c/V are not identical, they produce similar effects on the rates of reactions, so the terms are used somewhat interchangeably. For conditions where the internal pressure and the cohesion energy density are of equal magnitudes,

$$\frac{E_c}{V} = P_i = \frac{a}{V^2} \qquad (5.53)$$

For a mixture of components 1 and 2, the cohesion energy for the mixture, E_{cm}, will be given by

$$E_{cm} = \frac{a_m}{V_m} \tag{5.54}$$

If we represent the mole fraction of component 1 as X_1, then $X_2 = (1 - X_1)$ and the value of a_m is

$$a_m = X_1^2 a_1 + 2X_1(1-X_1)a_{12} + (1-X_1)^2 a_2 \tag{5.55}$$

where the interaction between components 1 and 2 is given by

$$a_{12} = (a_1 a_2)^{1/2} \tag{5.56}$$

which is known as the Berthelot geometric mean. If no change in volume occurs when the mixture is formed,

$$V_m = X_1 V_1 + (1 - X_1)V_2 \tag{5.57}$$

The cohesion energy of the mixture, E_{cm}, is given in terms of the cohesion energies of the two components by

$$E_{cm} = X_1 E_{c1} + (1 - X_1)E_{c2} \tag{5.58}$$

The change in cohesion energy when the mixture forms compared to the energy of the two components is

$$\Delta E_c = \frac{X_1(1-X_i)V_1 V_2}{X_1 V_1 + (1-X_1)V_2}\left(\frac{\sqrt{a_1}}{V_1} - \frac{\sqrt{a_2}}{V_2}\right)^2 \tag{5.59}$$

This equation is known as the Van Laar equation. Expressing the interaction of molecules of the liquids by means of van der Waals forces, the cohesion energy density is

$$\frac{E_c}{V} = \frac{a}{V^2} \tag{5.60}$$

Therefore,

$$\Delta E_c = \frac{X_1(1-X_1)V_1 V_2}{X_1 V_1 + (1-X_1)V_2}\left[\left(\frac{E_{c1}}{V_1}\right)^{1/2} - \left(\frac{E_{c2}}{V_2}\right)^{1/2}\right]^2 \tag{5.61}$$

which is known as the Hildebrand-Scatchard equation. Of course $(E_c/V)^{1/2}$ is the solubility parameter, δ, so we can write

$$\Delta E_c = \frac{X_1(1-X_1)V_1V_2}{X_1V_1+(1-X_1)V_2}[\delta_1 - \delta_2]^2 \qquad (5.62)$$

Thus, the difference in solubility parameters between the solvent and solute determines a great deal about the character of the solution. For example, water and carbon tetrachloride have cohesion energies which are approximately equal. However, the cohesion in water is due to dipole-dipole forces and hydrogen bonding while that in carbon tetrachloride is due to London dispersion forces. Mixing the two liquids results in the heat of mixing being positive because the strong interaction within the two components is not offset by forces that result between the polar and nonpolar molecules. Therefore, the two liquids do not mix.

We need now to consider the process of forming a solution from two components. We will represent the partial molar quantities of the *pure* components as G_i°, H_i°, and E_i° and those of the same components in *solution* as G_i, H_i, and E_i. The partial molar free energy, G_i, is related to that of the component in an *ideal* solution, G_i°, by the relationship

$$G_i - G_i^\circ = RT \ln a_i \qquad (5.63)$$

Using the analogous relationship for a *real* solution,

$$G_i - G_i^\circ = RT \ln \frac{a_i}{X_i} = RT \ln \gamma_i \qquad (5.64)$$

where γ is the activity coefficient. Therefore, because

$$\Delta G = \Delta H - T\Delta S$$

we can separate the free energy into the enthalpy and entropy components,

$$(H_i - H_i^\circ) - T(S_i - S_i^\circ) = RT \ln \gamma_i \qquad (5.65)$$

If molecular clustering does not occur and the orientation of each component is random in both the pure component and in the solution, the entropy of component i will be approximately the same in the solution as it is in the pure component. Therefore,

$$S_i = S_i^\circ \qquad (5.66)$$

The change in volume of mixing the liquids is almost zero so

$$\Delta H = \Delta E + \Delta(PV) \approx \Delta E \qquad (5.67)$$

and

$$\Delta E_i = E_i - E_i^\circ = RT \ln \gamma_i \qquad (5.68)$$

If the activity coefficient is approximately unity, the energy of one mole of component i is approximately the same in the solution as it is in the pure component. A relationship of this form is of great use in describing the thermodynamics of constituents of a solution.

5.5   EFFECTS OF SOLVENT COHESION ENERGY ON RATES

If the behavior of a reaction is considered in terms of volumes, the formation of the transition state can be viewed as the formation of a state having a different volume than that of the reactants. The change in volume can be written as

$$\Delta V^\ddagger = V^\ddagger - V_R \qquad (5.69)$$

where V_R is the volume of the reactants and $V^\ddagger$ is the volume of the transition state. It is important to note here that the internal pressure caused by the cohesion of the liquid results in an effect that is analogous to that produced by an external pressure (see Section 3.6). Accordingly, if the volume of activation is negative, the reaction will be enhanced when the solvent has a high internal pressure. Conversely, if the reaction has a positive volume of activation, the reaction will proceed faster in solvents having low internal pressure.

The effects of cohesion energy density or solubility parameter can be explained by considering a model in which cavities in the solution are altered as the reaction occurs. Cavity formation is hindered in solvents having large δ values. Moreover, the smaller volume species (either the reactants or the transition state) will be favored in such solvents. If the reactants exist in cavities having a larger total volume than that of the transition state, a solvent of high cohesion energy will favor the formation of the transition state.

In terms of an overall chemical reaction, the cohesion energy density can often be used as a predictor. If the products have greater cohesion energy density than the reactants, the process will be favored when a solvent of large δ is used. Conversely, if the reactants have high cohesion energy density, a solvent having a large δ retards the reaction. Predictably, if the reactants and products have similar cohesion energy densities, the δ value of the solvent will be relatively unimportant in its influence on the reaction. The cavities in a solution depend on the sizes of the species and the ability of the solvent to "compress" the cavity. Actually, if the solvent molecules are spherical, there will be free space in the pure solvent. For example, if a sphere is surrounded by eight others in a body-centered cubic arrangement, there is 32% free space in the structure.

When the interactions are of the "strong" dipole-dipole or hydrogen-bonding type, a solvent with a large δ value causes greater compression or electrostriction of the free space. The effects of using solvents having different solubility parameters on reaction rates will be explored in more detail in Section 5.10.

5.6 SOLVATION AND ITS EFFECTS ON RATES

The complexity of reactions in solution has already been described. However, many unimolecular reactions have rates in solutions which are about equal to those in the gas phase. The population of the transition state depends on the number of critical vibrational states populated, which is a function of temperature. The localization of the required energy in a vibrational mode for a bond to be broken is often rather independent of the environment of the molecule.

Generally, reacting molecules must come together and collide, form an activated complex and react, and allow the products to be removed by diffusion from the reaction zone. In viscous media, the collision frequency of the reactants to form the activated complex may limit the formation of the activated complex. Consider a process in which two solvated reactant molecules A and B must come together to form an activated complex. This process can be considered as requiring the close proximity of A and B (sometimes called a *collision complex*) followed by the actual formation of the critical configuration in space, which is the reactive transition state. This process can be shown as follows.

$$A(solv) + B(solv) \xrightarrow{\quad k_c \quad} AB(solv) \xrightarrow{\quad k_a \quad} [TS]^{\ddagger}(solv) \qquad \textbf{(5.70)}$$

In this scheme, AB(*solv*) is the collision complex of A and B while [TS]‡ (*solv*) is the actual transition state. We can characterize the rate of formation of the collision complex by the rate constant k_c and that of the formation of the activated state by k_a. The rate of diffusion of A and B in the solution determines k_c, and there is an activation energy associated with that process. In an approximate way, the activation energy can be considered as having a lower limit which is on the order of the activation energy for viscous flow. Such energies are generally lower than those for the formation of activated complexes during chemical reactions. Therefore, $k_c \gg k_a$ and the formation of the activated complex is rate-determining. In the case of viscous solvents and strong solvation of A and B, the formation of the collision complex of A and B may be rate-determining, in which case the reaction will be diffusion-controlled.

While diffusion-controlled reactions constitute a difficult special case, a general comparison of the behavior of gas phase reactions with those taking place in solution needs to be made. A problem with doing this is that few reactions that occur in the gas phase can be studied in solution under conditions that are otherwise the same with respect to temperature, concentration, etc. In some cases, even the products of the reaction may be different. The majority of studies on solvent effects have dealt with investigating the differences in kinetics of a reaction when different solvents are used rather than comparing gas phase reactions with those taking place in a solvent.

Let us consider the reaction between A and B that takes place in the gas phase and in some solvent to form the same products. We will write the processes in the two phases as follows.

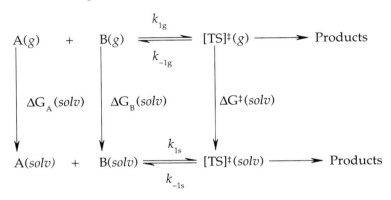

If the transition states formed under the two conditions are identical except for solvation and have equal probabilities for reaction, the rate of the reaction in each case will be determined by the concentration of the transition state.

$$\text{Rate} \propto [TS]^{\ddagger} \tag{5.71}$$

For the gas phase reaction,

$$R_g \propto [TS]^{\ddagger}_g = K_g^{\ddagger}[A]_g[B]_g \tag{5.72}$$

and for the reaction in solution,

$$R_s \propto [TS]^{\ddagger}_s = K_s^{\ddagger}[A]_s[B]_s \tag{5.73}$$

Therefore, when the concentrations of A and B are identical in the two phases, the ratio of the rates is given by the ratio of the equilibrium constants,

$$\frac{R_g}{R_s} = \frac{K_g^{\ddagger}}{K_s^{\ddagger}} \tag{5.74}$$

The equilibrium constants for the formation of the activated complexes in the two phases can be written in terms of the rate constants as

$$K_g^{\ddagger} = \frac{k_{1g}}{k_{-1g}} \quad \text{and} \quad K_s^{\ddagger} = \frac{k_{1s}}{k_{-1s}} \tag{5.75}$$

Consequently, making use of the principles shown in Chapter 2 by means of Eq. (2.113), we obtain

$$\frac{R_g}{R_s} = \frac{K_g^{\ddagger}}{K_s^{\ddagger}} = \frac{\exp(-\Delta G_g^{\ddagger}/RT)}{\exp(-\Delta G_s^{\ddagger}/RT)} \tag{5.76}$$

This equation can be written in logarithmic form as

$$\ln \frac{R_g}{R_s} = (\Delta G_s^{\ddagger} - \Delta G_g^{\ddagger})/RT \tag{5.77}$$

This equation shows that the difference in free energy of activation in the gas phase and in the solvent determines any difference in reaction rate. We can write Eq. (5.76) in terms of enthalpy and entropy of activation as

$$\frac{R_g}{R_s} = \frac{(\exp \Delta S_g^{\ddagger}/R)\,(\exp - \Delta H_g^{\ddagger}/RT)}{(\exp \Delta S_s^{\ddagger}/R)\,(\exp - \Delta H_s^{\ddagger}/RT)}$$

which can be written as

$$\frac{R_g}{R_s} = [\exp(\Delta S_g^{\ddagger} - \Delta S_s^{\ddagger})/R][\exp(\Delta H_s^{\ddagger} - \Delta H_g^{\ddagger})/RT] \tag{5.78}$$

It is readily apparent that when solvation effects on forming the transition state in solution are negligible compared to those on forming the transition state in the gas phase reaction, $\Delta S_g^{\ddagger} = \Delta S_s^{\ddagger}$ and $\Delta H_s^{\ddagger} = \Delta H_g^{\ddagger}$ so $R_g = R_s$ and the rate of reaction will be the same in the gas phase and in solution.

In a general way, we can see the effect of choice of solvent on a reaction by considering the free energy of activation. Figure 5.3 shows the cases that might be expected to arise when a reaction is studied in the gas phase and in four different solvents. In solvent 1, the reactants are strongly solvated so they reside at a lower free energy than they do in the gas phase. However, in this case the solvent is one that strongly solvates the transition state so it too resides at a lower free energy than it does in the gas phase, and by a greater amount than

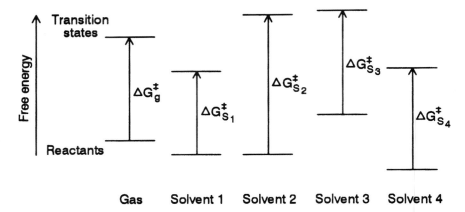

Figure 5.3   Effects of solvation of reactants and transition states on the free energy of activation. See text for explanation of the various cases.

do the reactants. Therefore, solvent 1 will increase the rate of reaction relative to that of the gas phase reaction because $\Delta G_{s1}^{\ddagger} < \Delta G_{g}^{\ddagger}$.

In solvent 2, solvation of the reactants leads to the reactants residing at a lower free energy, but the transition state is not solvated strongly and is destabilized compared to the gas phase transition state. Therefore, $\Delta G_{s2}^{\ddagger} > \Delta G_{g}^{\ddagger}$ and the reaction will proceed at a lower rate than it will in the gas phase. In solvent 3, neither reactants nor the transition state are well-solvated. In this case, the reactants and the transition state have higher free energies than they do for the gas phase reaction, but $\Delta G^{\ddagger}$ is unchanged. Therefore, the reaction should take place at about the same rate in this solvent as it does in the gas phase. Finally, in solvent 4, a solvent that strongly solvates both the transition state and the reactants, the rate should also be about the same as it is in the gas phase because $\Delta G_{g}^{\ddagger} \approx \Delta G_{s4}^{\ddagger}$.

The foregoing regards the effects of solvation on $\Delta G^{\ddagger}$. However, because

$$\Delta G^{\ddagger} = \Delta H^{\ddagger} - T\Delta S^{\ddagger}$$

it is apparent that an effect on $\Delta G^{\ddagger}$ could arise from a change in $\Delta H^{\ddagger}$ or $\Delta S^{\ddagger}$ (assuming that they do not change in a compensating manner; see Section 5.9). For example, when the reactions

$$CH_3Cl + N_3^- \longrightarrow CH_3N_3 + Cl^- \tag{5.79}$$

$$i-C_4H_9Br + N_3^- \longrightarrow i-C_4H_9N_3 + Br^- \tag{5.80}$$

were studied by Parker and coworkers (1968) in methanol and DMF (dimethyl formamide, $HCON(CH_3)_2$), the $\Delta H^{\ddagger}$ and $\Delta S^{\ddagger}$ values were found to reflect this difference. When $(\Delta H_M^\ddagger - \Delta H_{DMF}^\ddagger)/2.303RT$ and $(\Delta S_M^\ddagger - \Delta S_{DMF}^\ddagger)/R$ are compared for the reaction shown in Eq. (5.79) at 25°C, the values are 4.3 and 1.0, respectively. The value of $\Delta H_M^\ddagger - H_{DMF}^\ddagger$ being *positive* indicates that formation of the transition state is more difficult when methanol is the solvent than it is when the solvent is DMF. When expressed in the conventional way, $(\Delta H_M^\ddagger - \Delta H_{DMF}^\ddagger)$ = 24.5 kJ/mol while $(\Delta S_M^\ddagger - \Delta S_{DMF}^\ddagger)$ = 19 J/mol = 0.019 kJ/mol. It is apparent that the effect of changing solvents is due predominantly to the effect on $\Delta H^\ddagger$ and that entropy differences are only minor.

The origin of this solvent effect has been explored in the following way. The enthalpies of transfer, ΔH_{tr}, from water to dimethyl sulfoxide (DMSO), $(CH_3)_2SO$, of $(C_2H_5)_4NX$ (X = Cl, Br, or I) were determined and expressed relative to the value for the iodide compound. The free energies of changing solvent, ΔG_{tr}, of the corresponding silver compounds relative to AgI were also determined. By comparison of ΔH_{tr} for the $(C_2H_5)_4NX$ compounds with the ΔG_{tr} for the AgX compounds, the effects of cation cancel so the differences in ΔG_{tr} and ΔH_{tr} can be compared. The results given by Parker and coworkers (1968) are as follows for the anions (given as the anion: ΔG_{tr}(kJ/mol): ΔH_{tr} (kJ/mol)) Cl^- : 29 : 31; Br^-: 16 : 16; I^- : 0 : 0 (the reference). The fact that $\Delta G_{tr} \approx \Delta H_{tr}$ indicates that the entropy effects caused by changing solvent are negligible. Further, it is clear that the differences in behavior of ΔG_{tr} when changing solvents are primarily due, as far as the anions Cl^-, Br^-, and I^- are concerned, to differences in solvation enthalpies. In the case of these ions, the trend in solvation enthalpies from H_2O to DMSO is the one expected because Cl^- is smaller and a better electron donor in the formation of hydrogen bonds. Because Cl^- is a rather small and hard species, it is more strongly solvated in water, so ΔH_{tr} is more positive when Cl^- is transferred to DMSO. The bromide ion, being larger and softer, is not much more strongly solvated by water than it is by DMSO. One could expect that I^- might be about equally well solvated by the two solvents.

When a transition state is formed, the electronic charge distribution in a reactant molecule is changed. As a result, solvation factors are not static. A reactant may become a better electron donor or acceptor as the transition state forms, which may result in increased or decreased interactions with the solvent. Consequently, the Hammett σ constants that are obtained in one type of solvent (say a protic, polar solvent such as CH_3OH) may not apply quantitatively in a solvent such as DMSO or CH_3CN (see Section 5.8).

5.7 EFFECTS OF IONIC STRENGTH

When ions react in solution, their charges result in electrostatic forces that affect the kinetics of the reactions. We can see how this situation arises in the following way. If a reaction occurs between ions A^{Z_A} and B^{Z_B}, the transition state will be $[AB]^{ZA+ZB}$. The equilibrium constant for the formation of the transition state can be written in terms of activities as

$$K^{\ddagger} = \frac{a^{\ddagger}}{a_A a_B} = \frac{[TS]^{\ddagger}\gamma_{\ddagger}}{[A]\gamma_A[B]\gamma_B} \tag{5.81}$$

where the a is an activity and γ is an activity coefficient. From this equation, we obtain

$$[TS]^{\ddagger} = K^{\ddagger}\gamma_A\gamma_B[A][B]/\gamma_{\ddagger} \tag{5.82}$$

The rate of the reaction can be written as

$$-\frac{d[A]}{dt} = k[TS]^{\ddagger} = kK^{\ddagger}\frac{\gamma_A\gamma_B}{\gamma_{\ddagger}}[A][B] \tag{5.83}$$

$$-\frac{d[A]}{dt} = \frac{kT}{h}\cdot\frac{\gamma_A\gamma_B}{\gamma_{\ddagger}} \tag{5.84}$$

If the solution is sufficiently dilute so that the Debye-Huckel limiting law applies,

$$\log\gamma_i = -0.509Z_i^2 I^{1/2} \tag{5.85}$$

where Z is the charge on the species and I is the ionic strength of the solution. Therefore,

$$\log k = \log\frac{kT}{h}K^{\ddagger} + \log\frac{\gamma_A\gamma_B}{\gamma_{\ddagger}} \tag{5.86}$$

or

$$\log k = \log\frac{kT}{h}K^{\ddagger} + \log\gamma_A + \log\gamma_B - \log\gamma_{\ddagger} \tag{5.87}$$

Substituting for the log γ terms gives

$$\log k = \log\frac{kT}{h}K^{\ddagger} - 0.509(Z_A^2 + Z_B^2 - Z_{\ddagger}^2)I^{1/2} \tag{5.88}$$

The charge on the transition state $[TS]^{\ddagger}$ is $Z_A + Z_B$ so

$$Z_{\ddagger}^2 = (Z_A + Z_B)^2 = Z_A^2 + Z_B^2 + 2Z_A Z_B \tag{5.89}$$

Therefore, when this result is substituted in Eq. (5.88), we obtain

$$\log k = \log \frac{kT}{h} K^{\ddagger} + 1.018 Z_A Z_B I^{1/2} \qquad (5.90)$$

At constant temperature, the first term on the right-hand side of Eq. (5.90) is a constant, so it is apparent that a plot of log k versus $I^{1/2}$ should be linear. If one reactant is uncharged, Z_A or Z_B is equal to zero, and the ionic strength of the reaction medium should have little or no effect on the rate of the reaction. However, if A and B are both positive or both negative, the rate should increase linearly with $I^{1/2}$. If A and B are oppositely charged, the rate of the reaction should decrease linearly with $I^{1/2}$. In these cases, the slope of the plot of log k versus $I^{1/2}$ is directly proportional to the magnitude of $Z_A Z_B$. Observations on many reactions carried out in dilute solutions are in accord with these predictions.

The explanation for the observation when the product $Z_A Z_B$ is positive lies in the fact that when the ionic strength is high, the solvated ions change the dielectric behavior of the solution so that ions of like charge do not repel each other as greatly. This allows them to approach more closely, which causes an increase in collision frequency and an increased reaction rate. When the ions are of opposite charge, the change in the solvent characteristics causes a decreased attraction between the ions so that the rate of a reaction between them is decreased. Deviations from predicted behavior are common even when the solutions are quite dilute because the Debye-Huckel limiting law applies only to very dilute solutions. It should also be mentioned that ion pairing and complex formation can cause the relationship to be far from ideal.

For reactions that involve uncharged reactants, the ionic strength should be expected to have little effect on the reaction rate. If the reaction is one in which ions are consumed or generated, the overall ionic strength of the medium will change as the reaction progresses. Such a situation will alter the kinetic course of the reaction. In order to avoid this situation, one of two approaches must be used. First, the change in ionic strength can be determined and the results can be adjusted to compensate for the change. A more common approach is to carry out the reaction at a relatively high and essentially constant ionic strength by preparing a medium that contains a large concentration of an "inert" salt to provide an "ionic ballast." For many substitution reactions, the choice of salt is relatively easy since ions like ClO_4^-, NO_3^-, BF_4^-, or PF_6^- are not very good nucleophiles and do not compete with the entering group. If the reaction is one in which the electrophilic character of the cation is important, salts such as $R_4N^+Cl^-$ may be used. Obviously, some discretion must be exercised in the choice of "inert" salt in light of the reaction being studied. A realistic approach is to carry out the reaction making duplicate runs with different salts present at identical concentrations to ascertain that the salt in use is truly inert.

5.8   LINEAR FREE-ENERGY RELATIONSHIPS

The term *linear free-energy relationship* (LFER) applies to a variety of relationships between kinetic and thermodynamic quantities for both organic and inorganic reactions. About 70 years ago, J. N. Brønsted found a relationship between the dissociation constant of an acid, K_a, and its ability to function as a catalyst for a reaction that is catalyzed by an acid. The Brønsted relationship can be written as

$$k = CK_a^{\ n} \tag{5.91}$$

where k is a rate constant, K_a is the acid dissociation constant for the acid, and C and n are constants. Therefore,

$$\ln k = n \ln K_a + \ln C \tag{5.92}$$

or if common logarithms are used,

$$\log k = n \log K_a + \log C \tag{5.93}$$

Because

$$pK_a = -\log K_a \tag{5.94}$$

this equation can be written as

$$\log k = -n\, pK_a + \log C \tag{5.95}$$

This equation shows that a plot of $\log k$ versus pK_a should be linear. The anion, A^-, of the acid HA reacts with water,

$$A^- + H_2O \rightleftharpoons HA + OH^- \tag{5.96}$$

so we can write the equilibrium constant K_b as

$$K_b = \frac{[HA][OH^-]}{[A^-]} \tag{5.97}$$

Relationships for base catalysis similar to those shown in Eqs. (5.92) and (5.93) can be obtained.

$$\log k' = n' \log K_b + \log C' \tag{5.98}$$

or

$$\log k' = n' \log(K_w / K_a) + \log C' \tag{5.99}$$

In such cases, the reaction is dependent on the strength of the base, but it is inversely related to the strength of the parent acid.

Because the equilibrium constant for dissociation of an acid is related to the free energy change by

$$\Delta G_a = -RT \ln K_a \qquad (5.100)$$

substitution for $\ln K_a$ in Eq. (5.92) leads to

$$\ln k = -\frac{n\Delta G_a}{RT} + \ln C \qquad (5.101)$$

This equation shows that a linear free-energy relationship should exist between $\ln k$ for the acid-catalyzed reaction and ΔG_a for dissociation of the acid.

When two similar acids are considered, the rate constants will be given by k_1 and k_2 while the dissociation constants will be given by K_{a1} and K_{a2}. Then, subtracting the equation for the second acid from that for the first yields the equation

$$\ln k_1 - \ln k_2 = n(\ln K_{a1} - \ln K_{a2}) \qquad (5.102)$$

or

$$\ln \frac{k_1}{k_2} = n \ln \frac{K_{a1}}{K_{a2}} = nm \qquad (5.103)$$

where m is the logarithm of the ratio of the dissociation constants for the acids. As we shall see, when common logarithms are used and the constants are represented as ρ and σ, the relationship is known as the Hammett equation,

$$\log \frac{k}{k_o} = \rho\sigma \qquad (5.104)$$

where k_o is the rate constant for the reference acid. Numerous special cases of this type of equation exist where closely similar reactions are compared to a reference reaction. The rates of hydrolysis of alkyl halides have been correlated in this way.

The Hammett LFER relates the dissociation constants of substituted benzoic acids to that of benzoic acid itself. Described in 1937, the original relationship was developed to explain the electronic effects of substituents on m- and p-substituted benzoic acids. Then, the parameter, σ, was defined as

$$\sigma_m = \log \frac{K_{mx}}{K_o} \qquad (5.105)$$

$$\sigma_p = \log \frac{K_{px}}{K_o} \tag{5.106}$$

where K_o is the dissociation constant of benzoic acid and K_{mx} and K_{px} are the dissociation constants of the *m*- and *p*-XC_6H_4COOH acids. If the group X is electron-withdrawing, the acidity of the —COOH group is increased and σ is positive. Conversely, the σ values are negative for electron-releasing groups. When the dissociation constants for the acids $XC_6H_4CH_2COOH$ were studied, a linear relationship between $\log(K_x'/K_o')$ (K_o' is the dissociation constant of the reference, $C_6H_5CH_2COOH$) and the calculated σ values was obtained. Unlike the equations given above where the slope is unity, the constant slope was represented as ρ so that

$$\log \frac{K_x'}{K_o'} = \rho\sigma \tag{5.107}$$

When a series of reactions are studied in which the strength of the acid is rate-determining, the rates will be proportional to [H⁺] but this is in turn proportional to K_a. Therefore, the rate constants will be related by the equation

$$\log \frac{k_x}{k_o} = \rho\sigma \tag{5.108}$$

When other series of aromatic compounds are considered, the constants K_o and k_o refer to the reference unsubstituted acid. Eq. (5.108) shows that if $\rho > 1$, the rate or dissociation constant is enhanced by the electronic effects of substituent X to a greater extent than they are for the benzoic acids. On the other hand, if $\rho < 0$, the group X is *electron-releasing* and the rate (or dissociation) constant is increased by the group X. Finally, if $1 > \rho > 0$, the rate (or dissociation) constant is increased, but to a lesser extent than the benzoic acid is affected by the same substituent. While the major use of the Hammett relationship is in organic chemistry, a number of interesting correlations have been found for some inorganic reactions involving complexes as well.

The relationship

$$\log \frac{K_x}{K_o} = \rho\sigma \tag{5.109}$$

can be written as

$$\log K_x - \log K_o = \rho\sigma \tag{5.110}$$

and because

$$\Delta G = -RT \ln K = -2.303RT \log K$$

we find that

$$\log K = -\frac{\Delta G}{2.303RT} \tag{5.111}$$

Therefore,

$$\log K_x - \log K_o = \rho\sigma \tag{5.112}$$

and by substitution, we obtain

$$-\frac{\Delta G_x}{2.303RT} + \frac{\Delta G_o}{2.303RT} = \rho\sigma \tag{5.113}$$

which simplifies to

$$\Delta G_x = \Delta G_o - 2.303RT\rho\sigma = \Delta G_o - (\text{constant})\rho\sigma \tag{5.114}$$

which shows the *linear* relationship between the change in free energy and the product of $\rho\sigma$.

The LFER of Hammett is satisfactory only when the reactive site is sufficiently removed from the substituent so that steric factors do not enter into the rate-determining step. Also, if the reaction involves a series of substituents that greatly alter the way in which either the reactant or the transition state is solvated, the relationship may be less than satisfactory. It is perhaps wise to remember that the relationships are empirical in origin. This does not detract from their usefulness, neither is it any different for empirically determined rate laws. While the original Hammett LFER was applied to aromatic compounds, other studies have extended it to other types of compounds (e.g., aliphatics).

The approach of R. W. Taft is similar to that of Hammett, and the equation used is

$$\log \frac{k}{k_o} = \rho^* \sigma^* + \delta E_s \tag{5.115}$$

where σ^* is a constant related to polar substituent effects and ρ^*, as was the case for ρ, is a reactant constant, and δE_s is a steric energy term. When a given series of reactants is considered, δE_s is frequently considered to be zero since for any pair of similar species subtraction of two equations of the form of Eq. (5.115) would cause the term δE_s to cancel. Frequently, the Taft equation is written simply as

$$\log \frac{k}{k_o} = \rho^* \sigma^* \tag{5.116}$$

The Taft equation is essentially similar to the Hammett relationship but has constants appropriate to aliphatic and restricted aromatic materials.

5.9   THE COMPENSATION EFFECT

When a series of reactions involving similar reactants (e.g., a series of substituted molecules having different substituents in a particular position) is studied, it is possible to find that $\Delta G^{\ddagger}$ may show little variation for the series. This may be indicative of there being a relationship of the Hammett or Taft type. However, another explanation in some cases is the so-called *compensation effect.*

We can see how this situation might arise in a very simple way. As an extreme example, consider the solvation of ions that are present in reactions as the transition states TS_1 and TS_2. Suppose one has a charge of $+TS_1$ and the other has a charge that is $+TS_2$ where $TS_2 > TS_1$. In a polar solvent, TS_2 will be more strongly solvated than TS_1, so $\Delta H_2^{\ddagger}$ will be more negative than $\Delta H_1^{\ddagger}$. However, because this is true, the solvent in the vicinity of TS_2 will be more ordered than it is near TS_1, and $\Delta S_2^{\ddagger}$ will be more negative than $\Delta S_1^{\ddagger}$. The free energy of activation, $\Delta G^{\ddagger}$, is given by

$$\Delta G^{\ddagger} = \Delta H^{\ddagger} - T\Delta S^{\ddagger}$$

Therefore, if $\Delta H_2^{\ddagger}$ is more negative than $\Delta H_1^{\ddagger}$ and $\Delta S_2^{\ddagger}$ is more negative than $\Delta S_1^{\ddagger}$, it is possible that $\Delta G^{\ddagger}$ is approximately constant for the two cases. For a series of reactions, we might find that

$$\Delta H_i^{\ddagger} - T\Delta S_i^{\ddagger} = C \tag{5.117}$$

where C is a constant. Therefore, we can write

$$\Delta H_i^{\ddagger} = T\Delta S_i^{\ddagger} + C \tag{5.118}$$

and we should expect that a plot of $\Delta H^{\ddagger}$ versus $\Delta S^{\ddagger}$ should be linear with a slope of T. This temperature is sometimes called the *isokinetic temperature* in this *isokinetic relationship.* Figure 5.4 shows such a relationship for the reaction

$$[Cr(H_2O)_5OH]^{2+} + X^- \longrightarrow [Cr(H_2O)_5X]^{2+} + OH^- \tag{5.119}$$

where $X = Cl^-$, Br^-, I^-, SCN^-, etc. In this case, a reasonably good linear relationship results in spite of the fact that widely differing ligands were used. The mechanism of the substitution reactions involves the initial loss of OH^- followed by the entry of X into the coordination sphere.

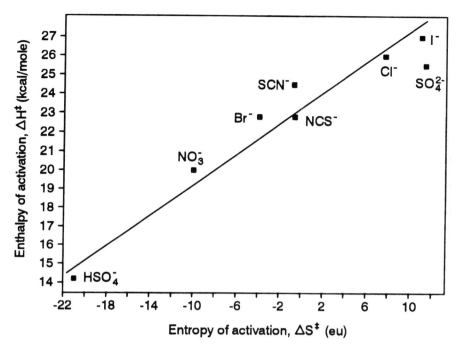

Figure 5.4   An isokinetic plot for the formation of $[Cr(H_2O)_5X]^{2+}$ by replacement of OH^-. (Constructed from the data given in D. Thusius, *Inorg. Chem.*, **1971**, *10*, 1106).

While we have interpreted the compensation effect in terms of transition states having different charges, there is no reason that transition states having different polarities could not behave similarly when the solvent is polar. Also, if the transition state involves the removal of charge separation and the solvent is nonpolar, more favorable solvation of the transition state would be expected on the basis of the hard-soft interaction principle. A compensation effect could result in this situation also. When a linear isokinetic relationship is obtained, it is usually taken as evidence for a common mechanism for the reactions.

5.10   SOME CORRELATIONS OF RATE WITH SOLUBILITY PARAMETER

The importance of solvent cohesion energy, as reflected by the solubility parameter, was described briefly in Section 5.5. Because the solubility parameter is such an important (and under-utilized) tool for explaining solvent effects on rates, we will describe here more of the details of a few studies. In a general way, solvents having large solubility parameters assist the formation of transition states in which there is high polarity or charge separation (high cohesion energy in the transition state). Conversely, solvents having large solubility parameters hinder the formation of transition states that have large, nonpolar structures.

A reaction which is widely cited as one in which solvent *polarity* plays a major role is the formation of quaternary ammonium salts by the reaction

$$R_3N + R'X \longrightarrow R'R_3N^+X^- \tag{5.120}$$

The transition state in this reaction is generally regarded as resembling the products (considerable charge separation, high cohesion energy). Accordingly, it would be logical to expect that the rate of the reaction would be enhanced by using solvents having large solubility parameters. One such reaction that has been studied by Kondo, et al. (1972) in a variety of solvents is

$$C_6H_5CH_2Br + C_5H_5N \longrightarrow C_5H_5NCH_2C_6H_5{}^+Br^- \tag{5.121}$$

In this case, it is found that the rate of the reaction increases when solvents having large solubility parameters are used. Moreover, the volume of activation for the reaction is negative in all solvents, but it is more negative in solvents having smaller solubility parameters.

The free energy of activation for a reaction having a compact, polar (or ionic) structure will be decreased by solvents having large solubility parameters. The equilibrium constant for the formation of the transition state is

$$K^{\ddagger} = \frac{k_1}{k_{-1}} \tag{5.122}$$

and the free energy of activation is related to $K^{\ddagger}$ by

$$\Delta G^{\ddagger} = -RT \ln K^{\ddagger} \tag{5.123}$$

Because the rate of the reaction will be proportional to the concentration of the activated complex, which is in turn proportional to $K^{\ddagger}$, we would expect that a plot of log k versus solubility parameter would be linear. We are assuming in this case that the decrease in the free energy of activation is directly proportional to the ability of the solvent to "force" the formation of a transition state. Figure 5.5 tests this relationship using the data given by Kondo, et al. (1972) for the reaction of benzylbromide with pyridine shown in Eq. (5.121). It can be seen that for most of the liquids the relationship is approximately correct.

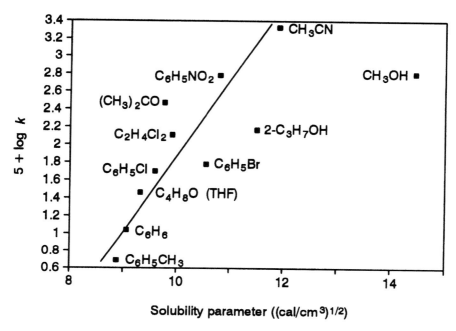

Figure 5.5 Relationship between the rate constant for the reaction shown in Eq. (5.121) and the solubility parameter of the solvent. (Constructed using the data of Kondo, *et al.* (1972)).

Furthermore, the ability of the solvent to solvate a transition state that has charge separation is related to the solubility parameter. As shown in Figure 5.6, the majority of the solvents give a satisfactory relationship between $\Delta V^{\ddagger}$ and δ in spite of the fact that widely differing solvents were used. However, two liquids, CH_3CN and CH_3OH, give data which fall far from the line. The solubility parameters for these liquids are large because of strong dipole-dipole forces (CH_3CN) and hydrogen bonding (CH_3OH). The fact that the volume of activation is more negative for solvents with lower cohesion energies is a reflection of the fact that these liquids have more loosely packed structures and that the reactants are much less constricted than they are in the transition state. If the transition state is approximately the same in volume when different solvents are used, the reactants must occupy a larger effective volume in the solvents of lower solubility parameter. The fact that the solvents CH_3CN and CH_3OH result in an abnormal volume of activation is probably due to the fact that these solvents have much more structure and the reactants already exist in small cavities as a result of the electrostriction of the solvent. These solvents are less compressible

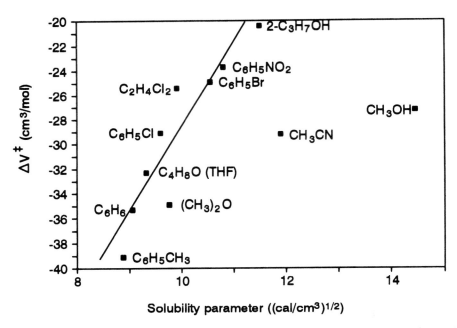

Figure 5.6   Relationship between the volume of activation for the reaction shown in Eq. (5.121) and the solubility parameter of the solvent. (Constructed using the data of Kondo, *et al.* (1972)).

and have already become tightly bound around the solutes. Consequently, there is a smaller volume change when the transition state forms.

We can examine the relationship between the solubility parameter of the solvent and the rate of a similar reaction by making use of the data given by Laidler (1965, p. 203) for the reaction

$$(C_2H_5)_3N + C_2H_5I \longrightarrow (C_2H_5)_4N^+I^- \qquad (5.124)$$

Figure 5.7 shows a plot of log k versus δ for this reaction carried out at 100° C. It is clear that the relationship is approximately linear, and as expected, the rate of the reaction increases with increasing solubility parameter of the solvent. As a general rule, we can conclude that *reactions which pass through transition states which have considerable polarity (or charge separation) induced will have rates that increase with increasing solubility parameter of the solvent.*

We should ask at this point what happens when reactions of a totally different type take place in solvents having different solubility parameters. Reactions in which the transition state is a large, essentially nonpolar species behave in exactly the opposite way with respect to the effects of solvent on the

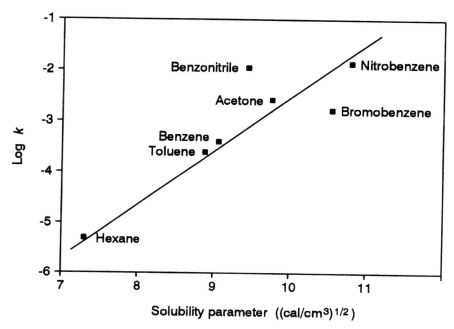

Figure 5.7 Relationship between the rate constants for the reaction between ethyl iodide and triethylamine and the solubility parameters of the solvents. (Constructed using data given by Laidler (1965)).

rate of reaction. One case of this type is the esterification that occurs when acetic anhydride reacts with ethanol at 50° C. Using the data given by Laidler (1965, p. 209), Figure 5.8 has been constructed. In this case, the rate of the reaction is seen clearly to decrease as the solubility parameter of the solvent increases. The formation of a large transition state having little or no charge separation from two smaller, polar molecules is hampered by solvents having high cohesion energy. In this case, a linear relationship also exists between log k and δ, but the slope is negative.

Parker and coworkers have investigated solvent effects on a variety of organic reactions. In one massive study on S_N2 reactions (Parker, et al., 1968), data are given for a large number of substitution reactions carried out in a wide range of solvents. Data for two of the numerous reactions studied were used to construct Figure 5.9, which shows the variation in log k as a function of the solubility parameter of the solvent. It is readily apparent that for the reactions

$$CH_3I + SCN^- \longrightarrow CH_3SCN + I^- \tag{5.125}$$

$$n - C_4H_9Br + N_3^- \longrightarrow n - C_4H_9N_3 + Br^- \tag{5.126}$$

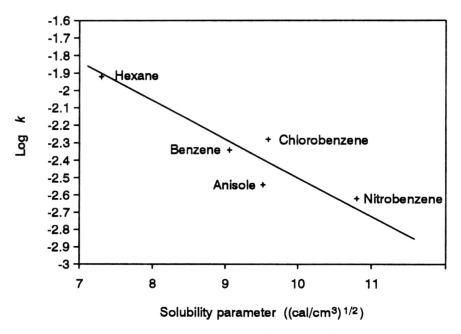

Figure 5.8   Dependence of the rate constants for the reaction between acetic anhydride and ethyl alcohol on the solubility parameters of the solvents. (Constructed using data given by Laidler (1965)).

the rate of substitution decreases more or less linearly with the solubility parameter of the solvent. In the first of these reactions, SCN^- is more strongly solvated by solvents that contain small, protic molecules (e.g., water or formamide) than is I^-. If a transition state such as

$$^-I \cdots \underset{\underset{H}{|}}{\overset{\overset{H}{\diagdown}\ \ \overset{H}{\diagup}}{C}} \cdots SCN^-$$

is formed, the charge is dissipated over a large structure so solvents that are essentially hard in character will not solvate the transition state as well as they will the SCN^-. On the other hand, solvents which are essentially soft in character (e.g., acetone, acetonitrile, or dimethylacetamide) will solvate the transition state more strongly than they will the reactants. As a result, the rate of the reaction will be greater in the softer solvents than it will in solvents that consist of small, protic molecules. Figure 5.9 shows that these observations are indeed borne out by the data for the reaction shown in Eq. (5.125).

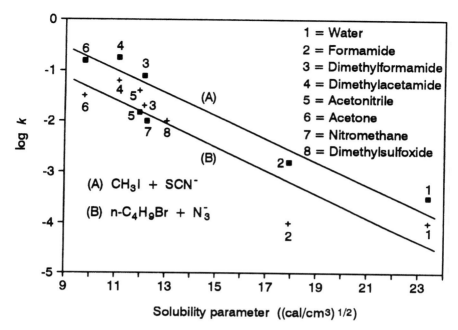

Figure 5.9 Relationship between log k and the solubility parameter of the solvent for nucleophilic substitution. (Constructed using the data of Parker, *et al.* (1968)).

Although the rates are lower for the reaction shown in Eq. (5.126) in each of the solvents tested, the same trend is seen for the reaction of *n*-butyl bromide with azide ion. The azide ion will be rather strongly solvated by solvents that have small, polar molecules while the bromide ion will be less well solvated. Also, the transition state will have the charge spread over a larger volume so that solvents which are composed of soft molecules will solvate the transition state better than those that consist of small, protic molecules.

While these cases illustrate the considerable role that the solubility parameter has in influencing the rates of reactions, especially in organic chemistry, space does not permit a consideration of the vast array of organic reaction types. Undoubtedly, reactions other than the substitution, esterification, and quaternization reactions described above are just as strongly influenced by the solvent. One of the best treatments of the broad area of solvent effects in organic chemistry is that by Leffler and Grunwald (1989), which contains an enormous amount of information. However, that source, as well as most others, does not really do justice to the general application of solubility parameters to explaining rates of reactions. In fact, the solubility parameter is of tremendous importance in predicting solution properties and other facets of liquid state science (Hildebrand and Scott, 1962, 1949).

REFERENCES FOR FURTHER READING

Alexander, R., Ko, F. C. F., Parker, A. J., Broxton, T. J. (1968) *J. Am. Chem. Soc.* *90*, 5049.

Bernasconi, G. F., Ed. (1986) *Part I, Investigations of Rates and Mechanisms of Reactions,* Vol. VI, in A. Weissberger, Ed., *Techniques of Chemistry,* Wiley, New York. Numerous chapters dealing with all aspects of kinetics in over 1000 pages.

Bernasconi, G. F., Ed. (1986) *Part II, Investigation of Elementary Reaction Steps in Solution and Fast Reaction Techniques,* Vol VI in A. Weissberger, Ed., *Techniques of Chemistry,* Wiley, New York. This book deals with many aspects of reactions in solution and solvent effects.

Cox, B. G. (1994) *Modern Liquid Phase Kinetics,* Oxford University Press, New York. A good survey of solution phase kinetics.

Dack, M. J. R., Ed. (1975) *Solutions and Solubilities,* Vol. VIII, in A. Weissberger, Ed., *Techniques of Chemistry,* Wiley, New York. Several chapters written by different authors deal with solution theory and reactions in solutions.

Ege, S. N. (1994) *Organic Chemistry: Structure and Reactivity,* 3d ed., D. C. Heath, Lexington, MA, p. 264.

Hildebrand, J., Scott, R. (1962) *Regular Solutions,* Prentice Hall, Englewood Cliffs, NJ. One of the most respected works on solution theory.

Hildebrand, J., Scott, R. (1949) *Solubility of Non-Electrolytes,* 3d ed., Reinhold, New York.

Kondo, Y., Ohnishi, M., Tokura, N. (1972) *Bull. Chem. Soc. Japan, 45,* 3579.

Laidler, K. J. (1965) *Chemical Kinetics,* McGraw-Hill, New York, p. 203, 209.

Leffler, J. E., Grunwald, E. (1989) *Rates and Equilibria of Organic Reactions,* Dover Publications, New York. A monumental classic that has appeared in reprint form. This book contains an enormous amount of material related to reactions in solution. A widely cited standard.

Lowry, T. H., Richardson, K. S. (1987) *Mechanism and Theory in Organic Chemistry,* 3d ed., Harper & Row, New York. Many basic ideas discussed in Chapter 2 with specific types of reactions and factors influencing them discussed in other chapters. Extensive treatment in linear free-energy relationships.

Wilkins, R. G. (1974) *The Study of Kinetics and Mechanism of Reactions of Transition Metal Complexes,* Allyn and Bacon, Boston. Detailed work on reactions of coordination compounds in solution.

PROBLEMS

1. For the linkage isomerization reactions of

$$[M(NH_3)_5ONO]^{2+} \longrightarrow [M(NH_3)_5NO_2]^{2+}$$

(where M = Co, Rh, or Ir) the activation parameters are as follows (Mares, M., Palmer, D. A., Kelm, H., *Inorg. Chim. Acta* **1978**, 27, 153).

	M = Co	M = Rh	M = Ir
$\Delta H^{\ddagger}$, kJ mol^{-1}	91.6 ± 0.8	80.2 ± 2.1	95.3 ± 1.3
$\Delta S^{\ddagger}$, J mol^{-1} K^{-1}	-17 ± 3	-33 ± 7	-11 ± 4

Test these data for an isokinetic relationship. Because the volumes of activation are -6.7 ± 0.4, -7.4 ± 0.4, and -5.9 ± 0.6 cm^3 mol^{-1}, what mechanism is suggested for these isomerization reactions?

2. The decomposition of diisobutyrylperoxide,

produces C_3H_6, CO_2, and $(CH_3)_2CHCOOH$. At 40°C, the following data were obtained (Walling, C., Waits, A. P., Milovanivic J., Pappiaonnou, C. G., *J. Am. Chem. Soc.* **1970**, 92, 4927).

Medium	$10^5 k$, sec^{-1}
Gas	1
Cyclohexane	4.70
Nujol	4.63
Benzene	23.8
Acetonitrile	68.1

(a) In light of these data, speculate on the nature of the transition state for this reaction. From this general knowledge of the transition state, propose a mechanism for the reaction. (b) While the solvent effects were originally explained partially in terms of solvent polarity, determine the relationship between solubility parameter of the solvent and k.

3. The hydrolysis of 2, 2-dimethyl-1, 3-dioxolane,

has been studied in mixtures of water and glycerol as the solvent (Schaleger, L. L., Richards, C. N., *J. Am. Chem. Soc.* **1970**, *92*, 5565). Activation parameters are as follows.

% Glycerol	$\Delta H^{\ddagger}$, kcal mol^{-1}	$\Delta S^{\ddagger}$, cal mol^{-1} K^{-1}
0	20.7	7.1
10	19.5	3.5
20	19.7	4.3
30	17.9	−1.6
40	17.6	−2.4

Analyze these data to determine whether a compensation effect is operative. If it is, determine the isokinetic temperature.

4. How should the following reactions depend on the ionic strength of the reaction medium?
   (a) $[Pt(NH_3)_3Br]^+ + CN^- \longrightarrow$
   (b) $[PtCl_4]^{2-} + CN^- \longrightarrow$
   (c) $[Pt(NH_3)_2Cl_2] + Cl^- \longrightarrow$

5. The solubility parameter for mixed solvents can be calculated using the equation

$$\delta_M = \sum_{i=1}^{n} X_i \delta_i$$

where X_i and δ_i are the mole fraction and solubility parameter for component *i*. The reaction

$$C_6H_5CH_2Cl + CN^- \longrightarrow C_6H_5CH_2CN + Cl^-$$

has been studied in mixtures of H_2O and dimethylformamide, DMF. Results obtained were as follows (Jobe, K.I., Westway, K.C., *Can. J. Chem.* **1993**, *71*, 1353).

Mole % DMF	$10^3 \times k$, M^{-1} s^{-1}
2.5	4.40
5.0	3.20
15.0	0.52
20.0	0.57

(a) Use the data shown in Table 5.2 to determine the solubility parameters of the mixed solvents. (b) Test the relationship between ln k and δ for the solvents. (c) Discuss the probable mechanism of the reaction in light of the effect of δ on the rate.

6. The hydrolysis of

takes place in solution. The rate varies with pH as follows: (Marecek, J. F., Griffith, D. L. *J. Am. Chem. Soc.* **1970**, *92*, 917).

pH	0.5*	1.48	2.01	2.97	4.32	6.03	9.59	10.0	10.8
$10^3 \times k_{obs}$	13*	6.64	5.90	5.83	5.04	5.05	8.31	11.3	23

*Estimated

Discuss what k_{obs} means in this case and describe a possible mechanism for the hydrolysis.

7. The compound 2, 2-azoisobutane undergoes decomposition in a 90:10 diphenyl ether : isoquinoline mixture with $\Delta H^{\ddagger}$ = 42.2 kcal/mol and $\Delta S^{\ddagger}$ = 16.2 cal/mol deg. In the gas phase, the values are 42.3 kcal/mol and 17.5 cal/mol deg, respectively. Explain what this signifies in terms of the transition state for the reaction. Speculate how the use of other solvents would likely affect the rate of the reaction.

8. The hydrolysis of

has been studied where Y can be one of several different substituents (Zaborsky, O. R., Kaiser, E. T. *J. Am. Chem. Soc.* **1970**, *92*, 860). The results obtained and the values of the *para* σ substituent constants (σ_p) are as follows.

Y	k, m M^{-1} sec^{-1}	σ_p
H	37.4	0.00
NH_2	13.6	−0.66
OCH_3	21.3	−0.27
CH_3	24.0	−0.17
Br	95.1	0.23
NO_2	1430	1.24

Compare the rate constants for the groups to that for hydrogen and test the relationship between $\log(k_H/k_Y)$ and σ_p to determine if a linear free-energy relationship exists. Provide an explanation for your results.

chapter 6

ENZYME CATALYSIS

Since the early study by Berzelius in 1835 to the preparation of the first crystalline enzyme in 1926 by Sumner and on to the present time, our knowledge of enzymes has grown at an increasing rate. Enzymes occur throughout the areas of biochemistry and physiology. These protein materials catalyze reactions, enabling them to take place at low temperatures under mild conditions. Many of these processes catalyzed by enzymes are of commercial importance. For example, β-amylase is used to cleave maltose (a disaccharide) units from starch in the preparation of corn syrup.

Enzymes are high molecular weight proteins that are linked by peptide bonds. In their catalytic behavior, they provide a lower energy pathway for a reaction to take place. It is generally accepted that this occurs by the binding of the reactant (the *substrate*) to form an enzyme-substrate complex, which renders the substrate more reactive in some particular way (Figure 6.1). We will describe the behavior of enzymes in more detail in the next section and then turn our attention to treating the kinetics of enzyme-catalyzed processes.

SUBSTRATE

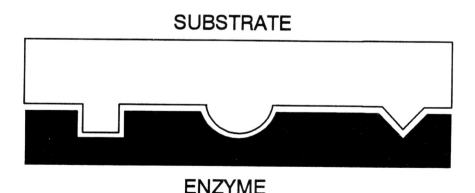

ENZYME

Figure 6.1 The "lock and key" model of the enzyme-substrate complex.

6.1 ENZYME ACTION

Enzymes frequently exhibit catalysis in very specific ways. In general, four types of behavior can be described.

1. *Absolute specificity.* In this type of behavior, the enzyme catalyzes a single reaction.
2. *Group specificity.* A reaction of only a single functional group is catalyzed by the enzyme.
3. *Linkage specificity.* In this case, the enzyme labilizes a specific type of bond independent of the overall substrate structure.
4. *Stereochemical specificity.* Some enzymes will induce reactions of only one isomer of a compound.

Some enzymes require the presence of another species before they are able to act as catalysts. Such additional species are called *cofactors,* and there are several types of cofactors known. The enzyme along with its cofactor is called the *holoenzyme,* while the protein portion alone is known as the *apoenzyme.*

Of the types of cofactors known, the most common are *coenzymes, prosthetic groups,* and *metal ions.* A coenzyme is some other organic material that is loosely attached to the protein enzyme (apoenzyme). If the organic compound is strongly attached to the apoenzyme, it is called a prosthetic group. Metal ions (e.g., Fe^{2+}, Ca^{2+}, Mg^{2+}, K^+, Cu^{2+}, etc.) may enhance enzyme activity by binding to the enzyme (complexation). On the other hand, some materials known as *inhibitors* reduce the activity of enzymes. Such cases may result from competitive inhibition in which some material can bind to the enzyme preventing its attachment to the substrate. In another mode of enzyme inhibition, noncompetitive inhibition, some material is present which binds to the enzyme changing its configuration so that it can no longer bind to the substrate effectively. Substrate inhibition may occur when a large excess of substrate is present which causes the equilibrium

$$E + S \rightleftharpoons ES \qquad (6.1)$$

to lie to the right so that most of the enzyme is effectively complexed.

Enzymes are protein materials and are, therefore, temperature-sensitive. The thermal stress caused by temperatures as low as $40\,^\circ C$ may be sufficient to cause denaturation of the protein causing a loss of catalytic activity. These changes may be due to slight configurational changes that require only small energy changes. The variation of rate of an enzyme-catalyzed reaction with temperature is shown in Figure 6.2. In general, the rate increases up to a point, then decreases at a higher temperature as denaturation of the enzyme takes

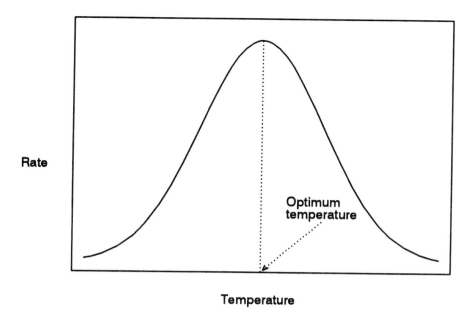

Figure 6.2 Dependence of enzyme-catalyzed reaction on temperature.

place. Accordingly, as shown in Figure 6.2, enzymes have optimum effectiveness in a rather narrow range of temperature.

Since enzymes are proteins, they contain acidic or basic sites. Basic sites may become protonated at higher [H$^+$] (low pH),

$$E + H^+ \rightleftharpoons H^+E \tag{6.2}$$

Acidic protons may be removed at high [OH$^-$] (high pH),

$$E + OH^- \rightleftharpoons H_2O + E^- \tag{6.3}$$

Either of these conditions can alter the effectiveness of an enzyme to catalyze a reaction. Therefore, a general relationship between enzyme activity and pH is shown in Figure 6.3. Most enzymes function best over a pH range of about 1.0 unit.

Enzymes are believed to function by complexing by what is called the "lock and key" fashion represented in Figure 6.1. The sites where the configuration of the enzyme and substrate match are called *active sites*. This type of interaction makes it easy to see why the action of enzymes is highly specific in many cases.

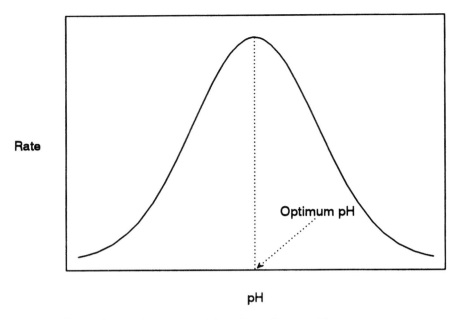

Figure 6.3   Dependence of enzyme-catalyzed reaction on pH.

6.2   KINETICS OF REACTIONS CATALYZED BY ENZYMES

Kinetic analysis of reactions catalyzed by enzymes is a difficult subject. However, many systems follow rather simple kinetic models which have been applied by many workers. While we will not treat some of the more esoteric and advanced topics associated with enzyme kinetics, a knowledge of the basic concepts is necessary for the student of chemical kinetics. We will now describe these concepts in sufficient detail for a book of this type.

6.2.1   Michaelis-Menten Analysis

When the concentration of substrate is varied over wide limits while the concentration of enzyme is held constant, the reaction rate increases until a certain concentration of the substrate is reached. This large concentration of substrate is sufficient to complex with all of the enzyme so any further increase in concentration of the substrate has no further effect on the rate. This situation is similar to that of OCl⁻ decomposition described in Section 1.2.3 (zero-order).

We can show the equilibrium in this case as

$$E+S \underset{k_{-1}}{\overset{k_1}{\rightleftharpoons}} ES \underset{k_{-2}}{\overset{k_2}{\rightleftharpoons}} P+E \tag{6.4}$$

In most cases, at least in the early stages, the concentration of the product is low so that the rate of the reverse reaction characterized by the rate constant k_{-2} can be neglected. Eventually, the rate of reaction ($k_2[ES]$) and decomposition ($k_{-1}[ES]$) of the enzyme/substrate complex equals the rate of its formation ($k_1[E][S]$) so that

$$k_1[E][S] = k_{-1}[ES] + k_2[ES] \tag{6.5}$$

The total enzyme concentration, $[E]_t$, is equal to the sum of the concentration of free enzyme, $[E]$, and that attached to the substrate, $[ES]$.

$$[E]_t = [E] + [ES] \tag{6.6}$$

Therefore,

$$[E] = [E]_t - [ES]$$

and by substituting this value in Eq. (6.5) we obtain

$$k_1([E]_t - [ES])[S] = k_{-1}[ES] + k_2[ES] \tag{6.7}$$

$$k_1[E]_t[S] - k_1[ES][S] = k_{-1}[ES] + k_2[ES] \tag{6.8}$$

which can be rearranged to give

$$k_1[E]_t[S] = k_{-1}[ES] + k_2[ES] + k_1[ES][S] \tag{6.9}$$

$$[ES] = \frac{k_1[E]_t[S]}{k_{-1} + k_2 + k_1[S]} \tag{6.10}$$

Since the rate of product formation, written as R, is $k_2[ES]$,

$$R = k_2[ES] = \frac{k_1 k_2 [E]_t [S]}{k_{-1} + k_2 + k_1[S]} \tag{6.11}$$

Dividing both numerator and denominator by k_1 gives

$$R = \frac{k_2[E]_t[S]}{\dfrac{k_{-1}+k_2}{k_1}+[S]}$$

(6.12)

or

$$R = \frac{k_2[E]_t[S]}{K_m+[S]}$$

(6.13)

This equation is known as the Michaelis-Menten equation and the constant $(k_{-1}+k_2)/k_1$ is called the Michaelis constant, K_m.

When [S] is large compared to K_m, the denominator of Eq.(6.12) is approximately [S] so that

$$R = \frac{k_2[E]_t[S]}{[S]}$$

(6.14)

and the rate under these conditions is the maximum rate, R_{max}.

$$R_{max} = k_2[E]_t$$

This case is equivalent to saying that all of the enzyme is bound to the substrate in complex form. Therefore,

$$R = \frac{R_{max}[S]}{K_m+[S]}$$

(6.15)

If the substrate concentration, [S], is equal to K_m,

$$R = \frac{R_{max}[S]}{[S]+[S]} = \frac{R_{max}}{2}$$

(6.16)

Therefore, while $K_m = (k_{-1}+k_2)/k_1$, it is also equal to the concentration of substrate when the reaction rate is half of its maximum value. This is shown in Figure 6.4. When the substrate concentration is much larger than K_m, [S] >> K_m,

$$R = \frac{k_2[E]_t[S]}{K_m+[S]} \approx \frac{k_2[E]_t[S]}{[S]} = k_2[E]_t$$

(6.17)

and the reaction follows a zero-order rate law with respect to substrate concentration. When K_m >> [S],

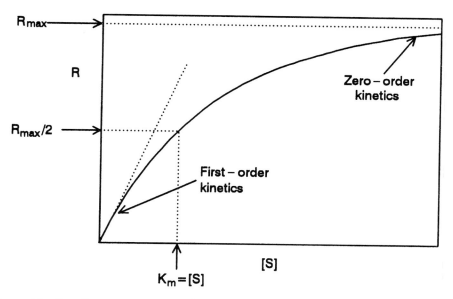

Figure 6.4   Reaction rate versus substrate concentration for a reaction following Michaelis-Menten kinetics.

$$R = \frac{k_2[E]_t[S]}{K_m + [S]} \approx \frac{k_2[E]_t[S]}{K_m} \tag{6.18}$$

so that the reaction follows a first-order rate law with respect to substrate concentration. These two regions are also shown in Figure 6.4.

The Michaelis constant is a fundamental characteristic of an enzyme. First, it gives the concentration of substrate necessary to bind to half of the available sites on the enzyme. Second, it gives an index of the relative binding affinity of the substrate to the enzyme active sites. Michaelis constants have been determined and tabulated for a large number of enzymes. Since K_m is a concentration, the values are usually expressed in **mM** units.

The Michaelis-Menten treatment of enzyme-catalyzed reactions bears a striking resemblance to the treatment of heterogeneous catalysis described in Section 4.6.1. Both treatments deal with the attachment of the reactant at active sites. Both also apply the steady-state approximation to the formation of "activated" reactant. Compare Figures 4.12 and 6.4 to see this similarity graphically and compare the forms of Eqs. (4.151) and (6.13) to see the similarity in rate laws.

While the Michaelis-Menten approach has been illustrated for the earlier and later portions of the plot shown in Figure 6.4, it is also possible to deal with Eq. (6.13) by direct integration.

$$-\frac{d[S]}{dt} = \frac{k_2[E]_t[S]}{K_m + [S]} \qquad (6.19)$$

Dividing both sides of this equation by $[S]/K_m + [S]$ gives

$$\frac{-\dfrac{d[S]}{dt}}{\dfrac{[S]}{K_m + [S]}} = k_2[E]_t$$

or

$$-\frac{d[S]}{dt} \cdot \frac{K_m + [S]}{[S]} = k_2[E]_t \qquad (6.20)$$

We can now write this equation as

$$-d[S]\left(\frac{K_m}{[S]} + 1\right) = k_2[E]_t\, dt \qquad (6.21)$$

or

$$-K_m \cdot \frac{d[S]}{[S]} - d[S] = k_2[E]_t\, dt \qquad (6.22)$$

Integrating this equation between limits $[S]_o$ at $t = 0$ and $[S]$ at time t, we obtain

$$K_m \ln\frac{[S]_o}{[S]} + ([S]_o - [S]) = k_2[E]_t\, t \qquad (6.23)$$

The first term on the left-hand side of this equation shows the first-order dependence of the rate, while the second term, $([S]_o - [S])$, shows the zero-order dependence. Figure 6.4 illustrates that the reaction starts off as a process that is first-order in substrate then shifts to a zero-order dependence.

The parallel between surface-catalyzed reactions described in Chapter 4 and enzyme-catalyzed processes has been mentioned. A comparison of Eq. (6.23) with Eq. (4.165),

$$\frac{1}{K}\ln\frac{P_{A,^o}}{P_A} + (P_{A,^o} - P_A) = kt$$

shows this relationship clearly. The surface-catalyzed reaction also shows the shift from first-order to zero-order kinetics. It should be pointed out that sometimes assuming similar mechanistic features results in equivalent mathematical models for widely differing branches of chemical kinetics. This is an intriguing feature of kinetic studies. For example, an equation found to be applicable to describing one type of solid state reaction is also applicable to describing fluorescence.

6.2.2 Lineweaver-Burk and Eadie Analyses

Kinetic analysis of enzyme-catalyzed reactions is conveniently carried out by writing Eq. (6.13) in the form

$$\frac{1}{R} = \frac{K_m + [S]}{k_2 [E]_t [S]} = \frac{K_m}{k_2 [E]_t [S]} + \frac{[S]}{k_2 [E]_t [S]} \tag{6.24}$$

$$\frac{1}{R} = \frac{K_m}{k_2 [E]_t [S]} + \frac{1}{k_2 [E]_t} \tag{6.25}$$

When $1/R$ is plotted versus the reciprocal of the substrate concentration, a straight line results, the slope of which is $K_m / k_2 [E]_t$ and the intercept is $1/k_2 [E]_t$. Such a plot, known as a Lineweaver-Burk or double-reciprocal plot, is shown in Figure 6.5. A disadvantage of this procedure is that most of the data are obtained

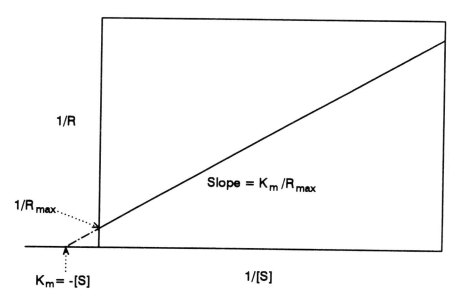

Figure 6.5 A Lineweaver-Burk plot for an enzyme-catalyzed reaction.

at relatively high substrate concentrations so that the extrapolation to low values of [S] may be somewhat inaccurate. If Eq. (6.25) is multiplied by [S], one obtains

$$\frac{[S]}{R} = \frac{[S]}{k_2[E]_t} + \frac{K_m}{k_2[E]_t} \tag{6.26}$$

A plot of [S]/R versus [S] is linear with a slope of $1/k_2[E]_t$, which is $1/R_{max}$ and intercept $K_m/k_2[E]_t$, which is K_m/R_{max}. This type of plot (known as a Hanes-Woolf plot) is shown in Figure 6.6.

Finally, Eq. (6.13) can be written as

$$R(K_m + [S]) = k_2[E]_t[S] \tag{6.27}$$

or

$$RK_m + R[S] = k_2[E]_t[S] \tag{6.28}$$

Dividing by [S] gives

$$\frac{RK_m}{[S]} + R = k_2[E]_t \tag{6.29}$$

In this case, a plot of R versus R/[S] is made, which has a slope of $-K_m$ and an intercept of $k_2[E]_t$. Such a plot, known as an Eadie-Hofstee plot, is shown in Figure 6.7. The methods based on Eqs. (6.26) and (6.29) are called single-reciprocal methods.

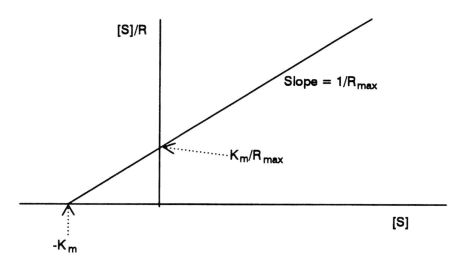

Figure 6.6 A Hanes-Woolf (single-reciprocal) plot.

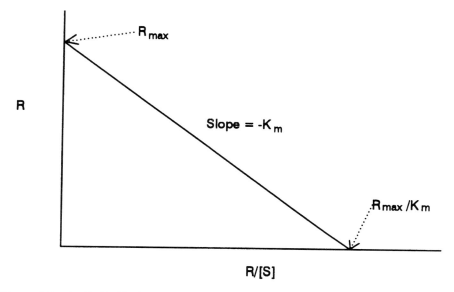

Figure 6.7   An Eadie-Hofstee (single-reciprocal) plot.

We have described some of the ways in which analyses of kinetic studies on reactions catalyzed by enzymes are carried out. This is an active area of research and many interesting and unusual aspects of these reactions are still being developed. Because of the space constraints of this book, it is not possible to describe the numerous cases here.

6.3   INHIBITION OF ENZYME ACTION

Although enzymes are protein materials having high molecular weights, some substrates are small molecules. For example, the decomposition of hydrogen peroxide (molecular weight 34) is catalyzed by the enzyme catalyase having a molecular weight of about 250,000. This enzyme lowers the activation energy for decomposition from 75 kJ/mol to about 8 kJ/mol. For many enzymes, the active site is localized to a small region of the much larger molecule. If some substance becomes bound to the active site of the enzyme, it loses part or all of its ability to function as a catalyst. The enzyme urease catalyzes the conversion of urea to NH_3 and CO_2. However, urease is strongly affected by metal ions such as Ag^+, Pb^{2+}, Hg^{2+}, or Pt^{2+}, which decrease the activity of the enzyme. Such substances are called *inhibitors.* It is the inhibition of peroxidase, catalyase, and cytochrome oxidase that causes HCN, H_2S, and azides to be toxic. Certain drugs act as inhibitors for the action of some enzymes. Consequently, it is essential that some discussion be presented on this important kinetic aspect of enzyme action. The three simplest kinetic models for enzyme inhibition are *competitive inhibition, noncompetitive inhibition, uncompetitive inhibition.*

6.3.1   Competitive Inhibition

We have earlier described the active site hypothesis for enzyme action. If a substance can bind at the enzyme active site, there is competition between the substrate, S, and the inhibitor, I, for the enzyme. The enzyme that is bound in a complex with the inhibitor, EI, is not available for binding with the substrate, so the effectiveness of the enzyme is diminished. The process can be represented as

$$E + S \underset{k_{-1}}{\overset{k_1}{\rightleftharpoons}} ES \underset{k_{-2}}{\overset{k_2}{\rightleftharpoons}} P + E \tag{6.30}$$

The concentration of EI is determined by the equilibrium

$$E + I \underset{k_{-3}}{\overset{k_3}{\rightleftharpoons}} EI \tag{6.31}$$

for which when E is substituted in the equilibrium constant expression gives

$$K = \frac{[EI]}{[E][I]} = \frac{k_3}{k_{-3}} = \frac{[EI]}{([E]_t - [ES] - [EI])[I]} \tag{6.32}$$

If we let K_i represent the equilibrium constant for dissociation of the EI complex, $K_i = 1/K$ and solving the resulting expression for [EI] gives

$$[EI] = \frac{[I]([E]_t - [ES])}{K_i + [I]} \tag{6.33}$$

For the complex ES, the change in concentration with time is the rate at which ES is formed minus the rate at which it dissociates. Therefore,

$$\frac{d[ES]}{dt} = k_1([E]_t - [ES] - [EI])[S] - k_{-1}[ES] - k_2[ES] = 0 \tag{6.34}$$

where $[E]_t$ is the total enzyme concentration. Substitution for [EI] from Eq. (6.33) yields upon rearranging

$$[ES] = \frac{[E]_t[S]K_i}{[S]K_i + K_mK_i + K_m[I]} \qquad (6.35)$$

Since the rate, R, is given by

$$R = k_2[ES]$$

the rate can now be expressed as

$$R = \frac{k_2[E]_t[S]K_i}{[S]K_i + K_mK_i + K_m[I]} \qquad (6.36)$$

The rate is maximum (R_{max}) when [S] is large, so under those conditions the rate reduces to $k_2[E]_t$. Therefore,

$$R = \frac{R_{max}[S]K_i}{[S]K_i + K_mK_i + K_m[I]} \qquad (6.37)$$

Writing this equation in terms of 1/rate gives

$$\frac{1}{R} = \frac{1}{R_{max}}\left(K_m + \frac{K_m[I]}{K_i}\right)\frac{1}{[S]} + \frac{1}{R_{max}} \qquad (6.38)$$

which is usually written in the form

$$\frac{1}{R} = \frac{K_m}{R_{max}}\left(1 + \frac{[I]}{K_i}\right)\frac{1}{[S]} + \frac{1}{R_{max}} \qquad (6.39)$$

This equation indicates that a graph of 1/R versus 1/[S] should be linear with a slope of $(K_m/R_{max})(1 + [I]/K_i)$ and an intercept of $1/R_{max}$. For different concentrations of inhibitor, different lines are obtained that are characteristic of [I]. The graphical application of this analysis is shown in Figure 6.8. When the concentration of inhibitor is varied and a family of lines such as that shown in Figure 6.8 is obtained, it is usually considered as diagnostic for a case of competitive inhibition.

6.3.2   Noncompetitive Inhibition

In this type of inhibition, the inhibitor is presumed not to bind to an active site of the enzyme, but rather to bind at some other site. This complex formation changes the conformation of the enzyme so that the substrate can not bind at the active site. The inhibition of urease by Ag^+, Pb^{2+}, or Hg^{2+} is believed to be

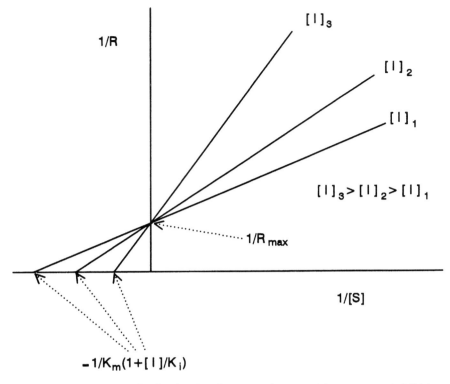

Figure 6.8   A Lineweaver-Burk plot for the case of competitive enzyme inhibition at three concentrations of inhibitor.

due to these ions binding to sulfhydryl (—SH) groups on the enzyme. For this type of action, we can write the equilibria

$$E + I \rightleftharpoons EI$$
$$ES + I \rightleftharpoons ESI$$

where both EI and ESI are inactive complexes of the enzyme. In this case, the equilibrium constants K_i^{EI} and K_i^{ESI} are the equilibrium constants for the dissociation of EI and ESI, respectively. Following a procedure similar to that in the last section, it is possible to derive the equation

$$\frac{1}{R} = \frac{K_m}{R_{max}}\left(1 + \frac{[I]}{K_i}\right)\frac{1}{[S]} + \frac{1}{R_{max}}\left(1 + \frac{[I]}{K_i}\right) \tag{6.40}$$

In this equation, K_i represents the combined effects of both K_i^{EI} and K_i^{ESI}. This equation indicates that a plot of $1/R$ versus $1/[S]$ would be linear with a slope of $(K_m/R_{max})(1 + [I]/K_i)$ and an intercept of $(1/R_{max})(1 + [I]/K_i)$. A different line will be obtained for each $[I]$ used. This situation is shown in Figure 6.9, which is usually taken as a diagnostic test for noncompetitive inhibition.

6.3.3   Uncompetitive Inhibition

This model describes a case in which the inhibitor combines reversibly with the complex ES after it forms. Further, it is assumed that this complex is so stable that it does not yield the product. If this were not so, this case would reduce to a special case of noncompetitive inhibition. The formation of the inactive complex is written as

$$ES + I \rightleftharpoons ESI \tag{6.41}$$

for which the formation constant can be written as

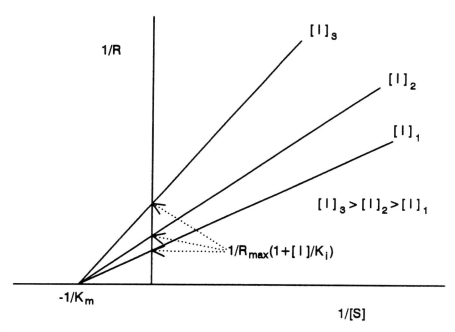

Figure 6.9   A Lineweaver-Burk plot for the case of noncompetitive enzyme inhibition at three concentrations of inhibitor.

$$K = \frac{[ESI]}{[ES][I]} \qquad (6.42)$$

Following the procedures used in the earlier cases, we can derive the equation

$$\frac{1}{R} = \frac{K_m}{R_{max}} \frac{1}{[S]} + \frac{1}{R_{max}}\left(1 + \frac{[I]}{K_i}\right) \qquad (6.43)$$

Equation (6.43) shows that a plot of $1/R$ versus $1/[S]$ will be linear. In fact, for a series of concentrations of inhibitor, a series of lines of identical slope (K_m/R_{max}) will be obtained, the intercepts of which will be related to [I] because the intercept is given by $(1/R_{max})(1 + [I]/K_i)$. This situation is shown graphically in Figure 6.10.

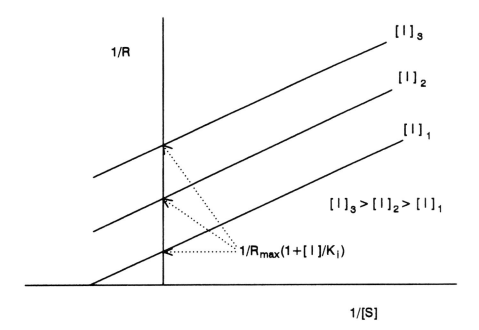

Figure 6.10   A Lineweaver-Burk plot for the case of uncompetitive enzyme inhibition at three concentrations of inhibitor.

6.4 ENZYME ACTIVATION BY METAL IONS

Earlier we mentioned the fact that certain enzymes exhibit enhanced activity by the presence of specific metal ions. For example, Mg^{2+} plays a role in phosphorylation reactions of adenosine triphosphate (ATP).

The transfer of a phosphate group is also assisted by a metal ion in the reaction

$$Glucose + ATP + Mg^{2+} \longrightarrow Glucose\text{-}6\text{-}phosphate + ADP + Mg^{2+}$$

A reaction scheme which shows the role of the metal ion in the enzyme activity is

$$E + M \underset{k_{-m}}{\overset{k_m}{\rightleftharpoons}} EM \tag{6.44}$$

$$EM + S \underset{k_{-1}}{\overset{k_1}{\rightleftharpoons}} EMS \overset{k_2}{\longrightarrow} E + P + M \tag{6.45}$$

where E is the enzyme, S is the substrate, M is the metal ion, and P is the product. It is possible to approach this mechanism by way of the steady-state approximation and obtain

$$R = \frac{R_{max}[M][S]}{K_m + K_m[M] + [S][M] + K_{EM}} \tag{6.46}$$

where K_m is the apparent Michaelis constant and K_{EM} is the equilibrium constant given by k_{-m}/k_m. For a given concentration of substrate, the rate varies with metal ion concentration and approaches a maximum value of $R_{max}[S]$.

6.5   REGULATORY ENZYMES

While the application of the Michaelis-Menten model has been illustrated, not all enzymes obey this model. Enzymes that catalyze reactions so that Michaelis-Menten kinetics is observed are called *nonregulatory enzymes*. *Regulatory* enzymes are those which are involved in a metabolic pathway, and they often give rate versus [S] plots that are sigmoidal. In fact, this feature is frequently the distinguishing feature of regulatory enzymes. Because the kinetics of regulatory enzymes does not follow the Michaelis-Menten model, the double reciprocal or Lineweaver-Burk plots will not be linear. Regulatory enzymes are sometimes compared by means of a parameter R_s, which is defined by the equation

$$R_s = \frac{\text{Substrate concentration at } 0.9\ R_{max}}{\text{Substrate concentration at } 0.1\ R_{max}} \tag{6.47}$$

For an enzyme that follows Michaelis-Menten kinetics, $R_s = 81$. For a regulatory enzyme that gives a sigmoidal curve, $R_s < 81$ if the enzyme is exhibiting *positive cooperativity*, a term which means that the substrate and enzyme bind in such a way that the rate increases to a *greater* extent with [S] than the Michaelis-Menten model predicts. Cases with $R_s > 81$ indicate *negative cooperativity* so that catalysis becomes *less* than is found in Michaelis-Menten kinetics. In these cases, kinetic analysis is usually carried out by means of the Hill equation,

$$\frac{R}{R_{max}} = \frac{[S]^n}{[S]^n + K'} \tag{6.48}$$

where K' is called the binding constant and n is the number of occupied sites on the enzyme where the substrate can bind. After taking the reciprocal of both sides, this equation can be rearranged to give

$$\frac{R_{max} - R}{R} = \frac{K'}{[S]^n} \tag{6.49}$$

or

$$\frac{R}{R_{max} - R} = \frac{[S]^n}{K'} \tag{6.50}$$

Taking the logarithm of both sides of this equation gives

$$\log \frac{R}{R_{max} - R} = n\ \log[S] - \log K' \tag{6.51}$$

Therefore, a plot of $\log(R/(R_{max} - R))$ versus $\log[S]$ should give a straight line having a slope of n and an intercept of $-\log K'$. Such a plot is known as a Hill plot. The terms *noncooperativity, negative cooperativity,* and *positive cooperativity* are applied to cases where $n = 1$, $n < 1$, and $n > 1$, respectively. When $R = R_{max}/2$,

$$\log \frac{R_{max}/2}{R_{max} - R_{max}/2} = \log\ 1 = 0 \tag{6.52}$$

and

$$n \log [S] = \log K' \tag{6.53}$$

Therefore,

$$[S]^n = K' \tag{6.54}$$

or

$$[S] = \sqrt[n]{K'} \tag{6.55}$$

It should be pointed out that sigmoidal rate plots are sometimes observed for reactions of solids. One of the rate laws used to model such reactions is the Prout-Tompkins equation, the left-hand side of which contains the function $\ln(\alpha/(1 - \alpha))$ where α is the fraction of the sample reacted (see Section 7.4). Compare the form of that equation with Eq. (6.51).

While we have treated a few cases of inhibition of enzyme action, several other possibilities are known and they have been mathematically described. A brief book such as this can provide only a survey of the topic of enzyme kinetics. While this vast, important, and rapidly growing branch of science can not be treated fully here, the introduction provided is sufficient for the nonspecialist in the field.

REFERENCES FOR FURTHER READING

Boyer, P. D., Lardy, H., and Myrback, K., Eds., (1959) *The Enzymes,* 2d ed., Academic Press, New York.

Kuchel, P. W., Ralston, G. B. (1988) *Schaum's Outline of Theory and Problems of Biochemistry,* Schaum's Outline Series, McGraw-Hill, New York, Chapter 9.

Laidler, K. J. (1958) *The Chemical Kinetics of Enzyme Action,* Oxford University Press, London.

Segel, I. H. (1974) *Biochemical Calculations,* 2d ed., Wiley, New York, Chapter 4.

Smith, E. L., Hill, R. L., Lehman, I. R., Lebkowitz, R. J., Handler, P., and White, A. (1983) *Principles of Biochemistry: General Aspects*, 7th ed., McGraw-Hill, New York, Chapter 10.

Sumner, J. B., Somers, G. F. (1953) *Chemistry and Methods of Enzymes*, 3d ed., Academic Press, New York, 1953, Chapter 1.

White, A., Handler, P., and Smith, E. (1973) *Principles of Biochemistry*, 5th ed., McGraw-Hill, New York, Chapter 11.

PROBLEMS

1. For an enzyme-catalyzed reaction, the following data were obtained.

[S] (**M**)	Rate (**M**$^{-1}$ min^{-1})
0.005	0.0143
0.010	0.0208
0.025	0.0294
0.050	0.0345
0.075	0.0360

   Using these data, determine the Michaelis constant for this enzyme system.

2. The reaction described in Question 1 was also carried out in the presence of an inhibitor, X, with the concentration of this inhibitor being 2.0×10^{-4} **M.** Under these conditions, the rate varied with substrate concentration as follows.

[S] (**M**)	Rate (**M**$^{-1}$ sec^{-1})
0.005	0.0080
0.010	0.0133
0.025	0.0222
0.050	0.0286
0.075	0.0323

   (a) Analyze these data to determine the type of inhibition that is caused by X.
   (b) Determine R_{max}.

3. The initial rate, V, of enzyme-catalyzed reaction varies with substrate concentration as follows:

[S] (**M**)	$10^6 \times$ Initial rate (**M** s^{-1})
0.020	0.585
0.004	0.495
0.002	0.392
0.001	0.312
0.00066	0.250

   Determine V_{max} and K_m for this reaction.

chapter 7

KINETICS OF REACTIONS IN THE SOLID STATE

One of the most important aspects of studying chemical dynamics is that of trying to deduce information about the mechanisms of reactions. This is an important component of any field of chemistry in which reactions are studied. It is no less the case for reactions involving solids. However, there are unusual aspects to solid state reactions that deserve special attention.

7.1 GENERAL CONSIDERATIONS

Because of the topics emphasized in the study of chemistry, our thinking about reactions is conditioned to events that occur during gas phase or solution phase reactions. For example, we are accustomed to thinking about a rate that involves *concentrations* of reactants raised to some appropriate power, the order of the reaction with respect to that component. The rate of a reaction in solution or the gas phase is expressed in terms of the change in concentration of some product or reactant with time. Thus, if

$$A \longrightarrow B \tag{7.1}$$

we can write

$$Rate = -d[A]/dt = d[B]/dt \tag{7.2}$$

If the reaction is first-order in A, the rate law can be written as

$$Rate = -d[A]/dt = k[A] \tag{7.3}$$

where k is the reaction rate constant. If the rate constant follows Arrhenius behavior,

$$k = Ae^{-E/RT} \tag{7.4}$$

where A is the pre-exponential (frequency) factor, E is the activation energy, and T is the temperature (K). Plotting $\ln(k)$ versus $1/T$ gives a straight line with a slope of $-E/R$ and an intercept of $\ln(A)$. As a general form, we can write

$$-d[\text{conc}]/dt = k\, f([\text{conc}]) \tag{7.5}$$

where [conc] is the concentration of a reactant and $f([\text{conc}])$ is some function of the concentration of this species. The concept of "order" is related to the molecularity of the reaction, the number of molecules forming the transition state. The mathematical treatment of such rate laws is covered in Chapters 1 and 2.

Reactions of materials in the solid state are strongly influenced by an enormous range of variables, and a complete treatment of this vast subject is beyond the scope of a book such as this. One factor that becomes apparent immediately in dealing with solid state reactions is that the rate can not be expressed in terms of concentrations. We can illustrate this using the following example. The first step in the decomposition of oxalates normally leads to the formation of carbonates. In the case of NiC_2O_4,

$$NiC_2O_4(s) \longrightarrow NiCO_3(s) + CO(g) \tag{7.6}$$

The density of NiC_2O_4 is 2.235 g/cm³. Accordingly, the concentration is (2.235 g/cm³ × 1000 cm³/l)/146.7 g/mole, which is 15.23 **M**. However, any particle of NiC_2O_4 has the same density and, hence, the same concentration. Thus, even if the particle changes size, its concentration does not change. The concentration in moles/liter is simply

$$\mathbf{M} = 1000d/(\text{MW}) \tag{7.7}$$

with d being the density (g/cm³) and MW is the formula weight (g/mol). For a particular solid phase, the concentration does not change, even as the particle reacts. The product phase represents an advancing phase boundary into the reactant phase but the "concentration" of the reactant phase does not change. It is well known, however, that the reactivity depends markedly on the configuration of the solid particles in some cases. Clearly, a different property is needed to express the rate of a reaction in the solid state.

The activation energy is obtained from the temperature dependence of the rate constant. For solid state reactions, there may not be an activated complex that is populated according to the Boltzmann Distribution Law. If we consider a few types of solid state reactions, we can see that there is no simple interpretation of k possible in some instances.

In a case where a gas reacts with a solid, such as tarnishing of a metal surface, the diffusion coefficient of the gas through the product layer determines the rate of the reaction. In some cases, it may be the rate of diffusion of the metal through a product layer of metal oxide that determines the rate of reaction. As the layer gets thicker, the rate decreases. A kinetic study of this process determines the activation energy for a *diffusion* process. It is not possible to attach the usual significance to the activation energy in terms of bond-breaking processes. For a reaction (studied electrochemically) such as

$$AgCl + Cu \longrightarrow Ag + CuCl \tag{7.8}$$

the temperature dependence of the rate gives a measure of the temperature effect on the conductivity of the products. These cases show that while the logarithmic function of the rate with $1/T$ is linear, the "activation energy" that results is likely to be for some process other than the breaking of chemical bonds.

Because we cannot represent the reaction rate in terms of concentrations, we must use some other approach. For reactions in the solid state, the fraction of the sample reacted, α, is frequently chosen as the reaction variable. Other possibilities include the thickness of the product layer, weight of product, or moles of product. It is immediately apparent that if α is the fraction of the sample reacted, $(1 - \alpha)$ is the fraction of the material remaining unreacted, and as we shall see, many rate laws are expressed in terms of $(1 - \alpha)$. The reaction rate can be expressed as *rate* = $d\alpha/dt$, and a reaction is complete when $\alpha = 1$. If we examine the behavior of α as a function of time for many reactions in the solid state, the general relationship is as shown in Figure 7.1.

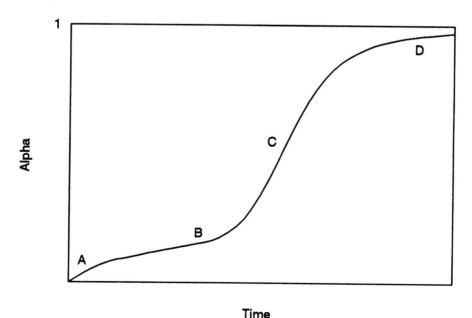

Figure 7.1   A general alpha versus time plot.

For many reactions of the type

$$R(s) \longrightarrow P(s) + G(g) \tag{7.9}$$

the most convenient way to determine the extent of reaction is by mass loss. Let us assume for the moment that such is the case here. Several regions can sometimes be detected for particular reactions. Region A represents the rapid evolution of adsorbed gases from the sample, resulting in a small initial mass loss. Region B represents an induction period where the rate is beginning to accelerate. In the region labeled C, the reaction is progressing at its maximum rate. Region D is called the *decay* region, and it represents a stage where the reaction is starting to slow down markedly as the reaction approaches completion.

Many reactions may not show all the steps illustrated in Figure 7.1. In fact, the majority do not show desorption of gas and although induction periods are fairly common, they are by no means universal. Consequently, it frequently is observed that the reaction starts off at its maximum rate then decreases thereafter. It is also sometimes observed that the reaction never progresses to $\alpha = 1$ for several reasons. One of the reasons is the retention of a gaseous product in the solid product and/or reactant phases. Retention is, therefore, regarded as the adsorption (or chemisorption) of a volatile product. The coalescence of particles due to sintering reduces surface area and increases the possibility that volatile products may be retained. The nature of the α versus time curve makes it very unlikely that a single rate law will describe the entire reaction.

7.2   FACTORS AFFECTING REACTION RATES IN SOLIDS

Because of the nature of reactions in the solid state, it is readily apparent that there are numerous factors that determine the reactivity of a particular sample. For example, if the reaction takes place on the surface of the solid, the particle size is important because the smaller the particles, the larger the surface area for a given volume of material. The nature of the surface itself is important because the reaction is determined by the *topochemistry*, the nature of the interface of reactant and product. Solid state reactions frequently depend on the formation and growth of active sites called *nuclei*. For some kinetic models, the geometric nature of the growth of these nuclei determines the form of the rate law for the reaction. Phase boundary advancement from these nuclei is another factor of importance. It is apparent that the rate of diffusion of material may play an important role in the behavior of the reaction. Lattice defects also are important because these defects promote diffusion and reactivity in general because they represent high energy centers in the solid. It has been observed that the reactivity of some solids increases dramatically in the temperature range where a phase transition occurs. Presumably, this is because at that temperature, the solid undergoes some type of crystal rearrangement. The mobility of lattice members at that temperature could promote reactivity in other processes than just the rearrangement. Finally, the historical treatment of the sample up to the time when its reaction is studied is important. The historical treatment may cause surface damage, cracks, pores, etc. Also, defects may have been "quenched in" by the preparative methods or they may have been annealed out. All of these factors make it quite likely that materials having the same apparent composition may react in drastically different ways or at drastically different rates.

7.3   RATE LAWS

There are approximately 20 rate laws that have found considerable application in the description of solid state processes. These have widely differing mathematical forms, and they are derived starting with certain models. Some of these will now be described to show how they are derived, and then we will summarize the most useful ones.

7.3.1   The Parabolic Rate Law

If we consider the reaction of oxygen with a solid (e.g., the oxidation of a particle of a metal), the oxide layer on the surface thickens as the reaction proceeds. Let x be the thickness of the layer. Then,

$$\text{Rate} = dx / dt \tag{7.10}$$

However, as x increases, the rate of reaction decreases because the oxygen must diffuse through the metal oxide layer. Therefore, the rate is proportional to $1/x$ so that

$$dx / dt = k(1 / x) \tag{7.11}$$

Then,

$$x \, dx = k \, dt \tag{7.12}$$

and integration yields

$$x^2 / 2 = kt \tag{7.13}$$

or

$$x^2 = 2kt \tag{7.14}$$

which can be written as

$$x = (2kt)^{1/2} \tag{7.15}$$

Because this equation has the form of an equation for a parabola, this rate law is referred to as the *parabolic rate law*. It is interesting to examine the units on k. If the thickness of the product layer is in cm and time is in sec, $k = cm^2/sec$. If we consider the weight of the layer (in g/cm^2), then $k = g^2 \ cm^{-4} \ sec^{-1}$. We can also express the amount of product layer as moles/cm^2 so $k = mole^2 \ cm^{-4} \ sec^{-1}$ in that case.

7.3.2   The First-Order Rate Law

Suppose we have a reaction of the type

$$R(s) \longrightarrow P(s) + G(g) \tag{7.16}$$

Let the amount of reactant R be represented by W. If the reaction is first-order,

$$-dW/dt = kW \tag{7.17}$$

If W_o represents the amount of reactant present initially and W represents the amount at some later time, the integrated rate law is

$$\ln(W_o/W) = kt \tag{7.18}$$

Now the fraction reacted, α, is given by

$$\alpha = (W_o - W)/W_o = 1 - (W/W_o) \tag{7.19}$$

Then $-W/W_o = \alpha - 1$ so that $W/W_o = 1 - \alpha$. Substituting this in Eq. (7.19) we obtain

$$-\ln(1-\alpha) = kt \tag{7.20}$$

The form of this equation shows that a plot of $-\ln(1 - \alpha)$ versus t would be linear with a slope of k and an intercept of zero. In fact, the intercept may not be zero due to an induction period or initial loss of adsorbed gases (see Figure 7.1). Therefore, a general form of the rate law is

$$-\ln(1-\alpha) = kt + C \tag{7.21}$$

where C is a constant. This and other rate laws based on the concept of reaction order (second, third, etc.) give α versus t curves which are deceleratory throughout. In other words, the reaction starts at its maximum rate and the rate decreases thereafter.

7.3.3   The Contracting Sphere Rate Law

We have described some of the difficulties that accompany kinetic interpretations of solid state reactions. In spite of this, a number of models can be described which represent essentially trial efforts for interpreting experimental results. We can expect, however, that detailed knowledge of a particular reaction is likely to be individualistic in its character. Having described the parabolic and first-order rate laws, we now look at some geometrical models. In a first simple case, we investigate the behavior of spherical solid particles. Suppose that a reaction involving a spherical solid particle of radius r takes place on the surface of the particle. For this particle, the volume is given by

$$V = (4/3)\pi r^3 \tag{7.22}$$

and the surface area is given by

$$S = 4\pi r^2 \tag{7.23}$$

If the reaction takes place on the surface of the sphere, the rate will be proportional to the surface area, S. However, assuming uniform density, the quantity of material present is proportional to V so that the volume of the particle is decreasing with time according to

$$-dV/dt = k_o S = k_o(4\pi r^2) \tag{7.24}$$

where k_o is a constant. But $r = (3V/4\pi)^{1/3}$ so that

$$-dV/dt = 4\pi k_o(3V/4\pi)^{2/3} = kV^{2/3} \tag{7.25}$$

where $k = 4k_o \pi(3/4\pi)^{2/3}$, a constant. The amount of material reacting is proportional to the volume so that this equation represents a process that involves "concentration" (actually volume) to the 2/3 power. Therefore, we say that this is a 2/3-"order" rate law. It should be apparent at this point that the concept of reaction order is of dubious meaning for solid state reactions.

Another way of interpreting this case is to substitute for V to obtain

$$-(d[(4/3)\pi r^3]/dr)(dr/dt) = -4\pi r^2 k_o = 4\pi r^2 (dr/dt) \tag{7.26}$$

or

$$-dr/dt = k_o \tag{7.27}$$

This expression is equivalent to saying that the radius of the reacting particle is decreasing at a constant rate.

The integrated form of the rate law is easily obtained.

$$V_o^{1/3} - V^{1/3} = kt/3 \tag{7.28}$$

Although this model is certainly a simplistic one, there are reactions known which do obey a 2/3-order rate expression.

Since the general kinetic treatment of solid state reactions can not be made in terms of concentrations, we need to put the integrated rate law above in the form containing α. In this case,

$$\alpha = (V_o - V)/V_o = 1 - (V/V_o) \tag{7.29}$$

Then

$$V/V_o = 1 - \alpha$$

or

$$(V/V_o)^{1/3} = (1-\alpha)^{1/3} \tag{7.30}$$

which can be written as

$$V^{1/3} = V_o^{1/3}(1-\alpha)^{1/3}$$

Substituting for $V^{1/3}$ in the rate law (Eq. 7.28),

$$V_o^{1/3} - V_o^{1/3}(1-\alpha)^{1/3} = kt/3 \tag{7.31}$$

Since the original volume of the particle, V_o, is a constant, this equation can be written as

$$1 - (1-\alpha)^{1/3} = k't \tag{7.32}$$

where $k' = k/3V_o^{1/3} = [4k_o\pi(3/4\pi)^{2/3}]/3V_o^{1/3}$. An equation of the same general form involving the expression $1 - (1-\alpha)^{1/3}$ results for other cases where a volume of material contracts in all three dimensions. For example, it can be shown that a cube which is reacting on the surface also leads to an equation of this general form, but the proof of that is left to the reader. A rate law of this general form is frequently referred to as a *contracting volume* rate law. The contracting volume model belongs to the deceleratory class of rate laws because the rate is maximum at the beginning of the reaction.

7.3.4 The Contracting Area Rate Law

Consider a particle of a solid in the shape of a cylindrical rod of radius r and length h as shown below.

Also suppose that the cylinder is very long compared to the radius, $h \gg r$. Therefore, the area of the ends can be considered as being insignificant compared to the area of the curved surface. We are, then, assuming that the length remains constant during the reaction. Since the amount of material is represented by the volume of the particle and the reaction occurs on the surface, we can write

$$-dV / dt = k_o S \tag{7.33}$$

But

$$S = \pi r^2 + 2\pi r h \tag{7.34}$$

Now the volume of a cylinder can be written as

$$V = \pi r^2 h$$

or

$$r = (V / \pi h)^{1/2}$$

Substituting for r in the rate equation gives

$$-dV / dt = 2\pi h k_o (V / \pi h)^{1/2} = 2k_o (\pi h)^{1/2} V^{1/2} \tag{7.35}$$

Therefore,

$$-dV / V^{1/2} = 2k_o (\pi h)^{1/2} dt \tag{7.36}$$

We can represent $2k_o (\pi h)^{1/2}$ as k and integrate from V_o at $t = 0$ to a volume of V at some later time.

$$-\int_{V_o}^{V} dV / V^{1/2} = k \int_0^t dt \tag{7.37}$$

so that

$$-2[V^{1/2} - V_o^{1/2}] = k't \tag{7.38}$$

Thus,

$$V_o^{1/2} - V^{1/2} = kt / 2 = k't \tag{7.39}$$

The reaction is, therefore, of 1/2-"order" in the volume (amount) of material present.

We can see how the radius changes with time as follows.

$$-dV / dt = -(dV / dr)(dr / dt) = 2\pi r k_o h \tag{7.40}$$

Now, from the relationship $V = \pi r^2 h$,

$$dV / dr = 2\pi rh$$

so

$$-2\pi rh(dr / dt) = 2\pi rhk_o$$

and

$$dr / dt = -k \tag{7.41}$$

That is, the radius of the particle decreases at a constant rate, exactly as it did in the case of the contracting sphere model. It is interesting that the contracting sphere gave a rate law that was 2/3-"order" and the present case, assuming the length (one dimension) remains constant, gives 1/2-"order." Thus, shrinking the particle in three dimensions leads to 2/3-"order" while shrinking the particle in two dimensions leads to 1/2-"order." These are general observations that are followed for particles of other geometry.

We can put the integrated rate equation in terms of α, the fraction reacted, because

$$\alpha = (V_o - V) / V_o \tag{7.42}$$

Therefore,

$$V / V_o = 1 - \alpha$$

and

$$(V / V_o)^{1/2} = (1 - \alpha)^{1/2} \tag{7.43}$$

which leads to

$$V^{1/2} = V_o^{1/2}(1 - \alpha)^{1/2}$$

Therefore,

$$V_o^{1/2} - V^{1/2} = kt / 2 \tag{7.44}$$

which gives by substitution for $V^{1/2}$,

$$V_o^{1/2} - V_o^{1/2}(1 - \alpha)^{1/2} = kt / 2 = V_o^{1/2}[1 - (1 - \alpha)^{1/2}]$$

Thus, the final equation can be written as

$$1 - (1 - \alpha)^{1/2} = kt / 2V_o^{1/2} = k't \tag{7.45}$$

A plot of $1 - (1 - \alpha)^{1/2}$ versus time will lead to a straight line if the reaction follows a 1/2-"order" rate law. This is one of the deceleratory rate laws known as the contracting area rate law.

A few words are in order at this point concerning the rate constant. If a plot is made of $1 - (1 - \alpha)^{1/2}$ versus time and a straight line results, the slope of that line will enable a calculation of k' to be made. If then the reaction is run at several temperatures, an Arrhenius plot can be made and an activation energy can be determined. This is assuming, of course, that the rate constant follows Arrhenius behavior, an assumption that can not always be made for solid state reactions. However, the constant k' has incorporated in it other factors. For this case, $k' = k/2V_o^{1/2}$ and $k = 2k_o(\pi h)^{1/2}$. Therefore, because the *measured* rate constant is not that of the original rate equation, the line made in the Arrhenius plot will be displaced upward or downward by a constant amount. While this does not affect the activation energy calculated from the slope, it *does* affect the intercept. Therefore, the Arrhenius plot can not be used directly to determine the pre-exponential factor from the intercept.

7.4   THE PROUT-TOMPKINS EQUATION

In Chapter 2, the effect of a product functioning as a catalyst was examined. It was shown that such a situation resulted in sigmoidal concentration versus time plots. In some cases, it appears that there is autocatalysis in the early stages of a reaction in the solid state. This is usually thought to be more important during the acceleratory period. An equation for autocatalysis is (see Section 2.6)

$$\ln \frac{\alpha}{(1-\alpha)} = kt + C \qquad (7.46)$$

This equation is based on a homogeneous reaction where the product can catalyze particles of the reactant. The derivation of this equation presented here follows closely that presented by Young (1966).

If N_o is the number of nuclei present at zero time, the change in number of nuclei, dN/dt, can be expressed as

$$\frac{dN}{dt} = k_o N_o + k_1 N - k_2 N \qquad (7.47)$$

The first two terms on the right-hand side of Eq. (7.47) give the number of nuclei originally present and the number produced by branching. The last term gives the loss of nuclei, which results when nuclei are terminated. Termination occurs when a spreading nucleus encounters product and can not continue to spread as a reaction site. After the original nucleation sites are spent, Eq. (7.47) will become

$$\frac{dN}{dt} = (k_1 - k_2)N \tag{7.48}$$

For linear nuclei as the reactant sites, the fraction of the sample reacted will vary with the number of nuclei as

$$\frac{d\alpha}{dt} = kN \tag{7.49}$$

In order to arrive at a final relationship for α as a function of time, it is necessary to obtain a relationship between α and the constants of Eq. (7.48). For a symmetrical sigmoidal curve, there will be an inflection point, α_i, at 0.5. At the inflection point, $d\alpha/dt$ changes from positive to negative and because the slope of the first derivative at the inflection point is zero, $k_1 = k_2$. Therefore, at this point, $k_2 = k_1\alpha/\alpha_i$. Substituting this result for k_2 in Eq. (7.48) we find

$$\frac{dN}{dt} = k_1N - k_2N = k_1N - \frac{\alpha}{\alpha_i}k_1N \tag{7.50}$$

which can be written as

$$\frac{dN}{dt} = k_1N\left(1 - \frac{\alpha}{\alpha_i}\right) \tag{7.51}$$

From Eq. (7.49),

$$N = \frac{1}{k}\frac{d\alpha}{dt} \tag{7.52}$$

Substituting this result in Eq. (7.51) we obtain

$$\frac{dN}{dt} = \frac{k_1}{k}\frac{d\alpha}{dt}\left(1 - \frac{\alpha}{\alpha_i}\right) \tag{7.53}$$

By removing dt, we can write this equation as

$$\frac{dN}{d\alpha} = \frac{k_1}{k}\left(1 - \frac{\alpha}{\alpha_i}\right) \tag{7.54}$$

or

$$dN = \frac{k_1}{k}\left(1 - \frac{\alpha}{\alpha_i}\right)d\alpha = \frac{k_1 d\alpha}{k} - \frac{k_1}{k}\frac{\alpha d\alpha}{\alpha_i} \tag{7.55}$$

This equation can be integrated to give the relationship between α and the number of nuclei,

$$N = \frac{k_1}{k} \alpha \left(1 - \frac{\alpha}{2\alpha_i}\right) \tag{7.56}$$

and if $\alpha_i = 0.5$,

$$\frac{d\alpha}{dt} = kN = k\frac{k_1}{k}\alpha(1 - \alpha) = k_1\alpha(1 - \alpha) \tag{7.57}$$

Therefore,

$$d\alpha = k_1\alpha(1 - \alpha)dt$$

or

$$\frac{d\alpha}{\alpha(1 - \alpha)} = k_1 dt \tag{7.58}$$

The integral on the left-hand side is of the form

$$\int \frac{dx}{x(ax + b)} = \frac{1}{b}\ln\frac{x}{(ax + b)}$$

so that the rate law can be written as

$$\ln\frac{\alpha}{(1 - \alpha)} = k_1 t + C \tag{7.59}$$

This rate law describes a process with linear branching chain nuclei which can be terminated when they reach the product phase. It is generally used to analyze the acceleratory period of reactions (typically up to $\alpha = 0.3$ or so). An equation of this type was used by Prout and Tompkins to study the decomposition of $KMnO_4$, and it has also been applied to the decomposition of silver oxide.

7.5   RATE LAWS BASED ON NUCLEATION

Many chemical reactions follow rate laws that are based on the formation of nuclei. Such active sites have been observed microscopically in some cases, and the phenomenon is well established. While they will not be described in detail here, several other processes have nucleation as part of at least their early stages. For example, crystal growth has been modeled by this type of rate law. Condensation of droplets also involves a process of nucleation. Consequently, kinetics of a wide variety of transformations obey rate laws based on nucleation, and even some polymerization reactions follow a form of the nucleation rate law.

The general form of the rate law that is used to describe nucleation processes is the Avrami (or Avrami-Erofeev) rate law,

$$\alpha = 1 - \exp(-kt^n) \tag{7.60}$$

which is written in logarithmic form as

$$[-\ln(1-\alpha)]^{1/n} = kt \tag{7.61}$$

In this rate law, n is the *index* and it usually has values of 1.5, 2, 3, or 4. In particular, the A1.5 rate law has been used to describe crystallization processes. The rate laws having $n = 2$ and $n = 3$ are associated with two- and three-dimensional growth of nuclei, respectively.

Nuclei may be present initially or they may grow in by a process that is usually considered to be first-order. The derivation of the Avrami rate law can be accomplished by several means (Young, 1966), all of them rather complicated. In general, assumptions are made regarding the rate of change of nuclei and the volume swept out by them as they react. When the volume of all nuclei is considered along with the volume change of nuclei as they react, it is possible to derive the equation

$$-\log(1-\alpha) = C\left(\exp(-kt) - 1 + kt - \frac{(kt)^2}{2!} + \frac{(kt)^3}{3!} \right) \tag{7.62}$$

(where C is a constant representing a collection of constants, involving among other things, N_o), which is the most general form of the random nucleation rate law. At longer times (the decay region of the α versus time curve), the term $(kt)^3/3!$ dominates so that

$$-\log(1-\alpha) = C'k't^3 \tag{7.63}$$

which can be written as

$$\alpha = 1 - \exp(-C'k't^3) \tag{7.64}$$

It can also be shown that the early stages of the reaction have α varying approximately as t^4. The general equation

$$\alpha = 1 - \exp(-kt^n) \tag{7.65}$$

can also be derived from certain geometric models for the decomposition of solid crystalline hydrates (Young, 1966).

Table 7.1 shows data for a reaction which follows an Avrami-Erofeev rate law with an index of 2 when the rate constant is 0.025 min^{-1}. These data yield the sigmoidal plot shown in Figure 7.2.

Table 7.1   Alpha versus time data for a reaction following an Avrami-Erofeev rate law with $n = 2$ and $k = 0.025$ min^{-1}.

Time (min)	α	Time (min)	α
0	0.000	55	0.849
5	0.016	60	0.895
10	0.061	65	0.929
15	0.131	70	0.953
20	0.221	75	0.970
25	0.323	80	0.981
30	0.430	85	0.989
35	0.535	90	0.994
40	0.632	95	0.996
45	0.718	100	0.998
50	0.790		

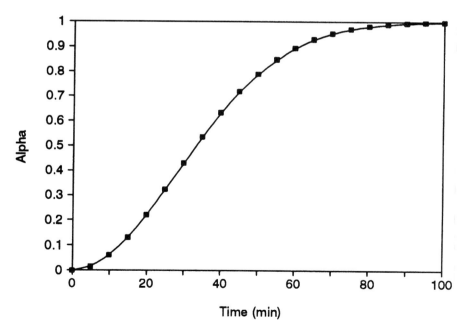

Figure 7.2   A plot of alpha versus time for the data shown in Table 7.1.

This type of curve is typical of processes following a nucleation rate law. In Chapter 2, it was shown that a sigmoidal plot results from autocatalysis, but for reactions in the solid state, such plots are more likely to indicate that the reaction rate is controlled as a result of a nucleation process.

Many reactions in the solid state follow a rate law of this Avrami-Erofeev type. For example, the dehydration of $CuSO_4 \cdot 5H_2O$ is a process where the first two steps are

$$CuSO_4 \cdot 5H_2O(s) \longrightarrow CuSO_4 \cdot 3H_2O(s) + 2H_2O(g) \qquad \textbf{(7.66)}$$

$$CuSO_4 \cdot 3H_2O(s) \longrightarrow CuSO_4 \cdot H_2O(s) + 2H_2O(g) \qquad \textbf{(7.67)}$$

These reactions take place in the temperature range of 47 to 63°C and 70.5 to 86°C, respectively. Both reactions appear to follow an Avrami-Erofeev rate law with an index of 2 over a range of α from 0.1 to 0.9 (Ng, *et al*, 1978).

For a rate law of the form $[-\ln(1 - \alpha)]^{1/n} = kt$, the most obvious way to evaluate the constants n and k is to take the logarithm of both sides of the equation so that one obtains

$$(1/n) \ln [-\ln(1 - \alpha)] = \ln(kt) = \ln(k) + \ln(t) \qquad \textbf{(7.68)}$$

Now a plot of $\ln[-\ln(1 - \alpha)]$ versus $\ln(t)$ should give a straight line with a slope of n and an intercept of $n[\ln(k)]$ when the correct value of n is used. In practice, it is preferable to plot the function $[\ln(1/(1 - \alpha))]^{1/n}$ versus t. This is because the form involving the ln ln plot is insensitive due to the nature of that function.

7.6   KINETIC STUDIES

The rate laws studied up to this point involve a large number of forms. Those and a few others are listed in Table 7.2. It is readily apparent that several of the rate laws are of very similar mathematical form. Consequently, applying these equations to actual data may result in more than one of the equations giving about equally good fit to the data. It is frequently the case that it is virtually impossible to say with certainty which rate law is correct if all one has to go on is the data for α as a function of time. In these cases, it is helpful to have data from several different experiments so that errors in the fraction of the reaction complete do not make it impossible to identify the correct rate law. In most cases, it is necessary to have data from several runs in order to establish the correct rate law.

Figure 7.3 shows Avrami-Erofeev rate plots of the data shown in Table 7.1. In this case, the indices of reaction, n, of 2, 3, and 1.5 were used to generate the $[-\ln(1 - \alpha)]^{1/n}$ functions used in the plots. It is clear that if only a few data were

Table 7.2   Classification of rate laws having the form f(α) = kt
for solid state reactions.

1. Acceleratory alpha-time curves	
P1 power law	$\alpha^{1/n}$
E1 exponential law	$\ln \alpha$
2. Sigmoidal alpha-time curves	
A2 Avrami-Erofeev (index 2)	$[-\ln(1 - \alpha)]^{1/2}$
A3 Avrami-Erofeev (index 3)	$[-\ln(1 - \alpha)]^{1/3}$
A4 Avrami-Erofeev (index 4)	$[-\ln(1 - \alpha)]^{1/4}$
A1.5 Avrami-Erofeev (index 1.5)	$[-\ln(1 - \alpha)]^{2/3}$
B1 Prout-Tompkins	$\ln[\alpha/(1 - \alpha)]$
3. Deceleratory alpha-time curves	
A. Based on geometrical models	
R2 contracting area	$1 - (1 - \alpha)^{1/2}$
R3 contracting volume	$1 - (1 - \alpha)^{1/3}$
B. Diffusion controlled mechanisms	
D1 one-dimensional diffusion	α^2
D2 two-dimensional diffusion	$(1 - \alpha)\ln(1 - \alpha) + \alpha$
D3 three-dimensional diffusion	$[1 - (1 - \alpha)^{1/3}]^2$
D4 Ginstling-Brounshtein	$1 - 2\alpha/3 - (1 - \alpha)^{2/3}$
C. Based on reaction order	
F1 first-order	$-\ln(1 - \alpha)$
F2 second-order	$1/(1 - \alpha)$
F3 third-order	$[1/(1 - \alpha)]^2$

available and if they were subject to substantial errors, as they frequently are, it might be difficult to determine the correct index of reaction. Also, some of the other functions shown in Table 7.2 might fit the data about equally well. In the case shown, the data were calculated assuming that $k = 0.025$ min^{-1} and the entire range of α values was used. If a smaller range of α is considered and if the data are in error due to experimental conditions, the selection of the correct rate law is not a trivial problem.

As was illustrated in Chapter 1, errors in data can make it difficult to distinguish the correct rate law, especially when the reaction is followed only to 50 or 60% completion. For solid state reactions, it may be virtually impossible to assign a unique rate law under these conditions because of the mathematical similarity of the rate laws. The general rule for determining the rate law in kinetic studies is to follow the reaction over several half-lives (see Chapter 1). However, this is almost never possible for a solid state reaction and even if it were, the rate law might be different in different stages of the reaction (see

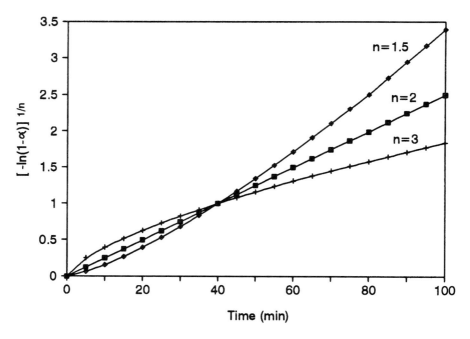

Figure 7.3   Avrami-Erofeev plots of the data shown in Table 7.1.

Section 7.1). Data points in the range $0.1 < \alpha < 0.9$ are generally chosen to avoid any induction period and decay region unless these regions are specifically under study. As a result of these factors, it is generally necessary to make a large number of kinetic runs to try to determine a rate law which is applicable to the reaction. This will be illustrated in the next section.

In Chapter 8, we will see that most of the rate laws shown in Table 7.2 can be put in the form of a composite rate law involving three exponents. We will also discuss the difficulties in determining these exponents from (α, t) data. While the discussion up to this point has set forth the basic ideas of solid state reactions, their application to specific studies has not been shown. We will now examine the results of three case studies to illustrate the type of information that kinetic studies on reactions in the solid state produce.

7.6.1   The Deaquation-Anation of $[Co(NH_3)_5H_2O]Cl_3$

When aquapentamminecobalt(III) chloride and similar compounds are heated, the water is lost and an anion enters the coordination sphere of the metal.

$$[Co(NH_3)_5H_2O]Cl_3\,(s) \longrightarrow [Co(NH_3)_5Cl]Cl_2\,(s) + H_2O(g) \quad \textbf{(7.69)}$$

However, when other anions (e.g., Br^-, NO_3^-, or SCN^-) are present, the kinetics of the reaction is altered. For example, activation energies of 79, 105, and 139 kJ mol^{-1} have been reported for the Cl^-, Br^-, and NO_3^- compounds, respectively. This behavior, an anion effect, has led various workers to postulate a mechanism like that shown in Scheme I.

Scheme I

$$[Co(NH_3)_5H_2O]Cl_3 \overset{\text{slow}}{\rightleftharpoons} [Co(NH_3)_5H_2OCl]Cl_2$$

$$[Co(NH_3)_5H_2OCl]Cl_2 \xrightarrow{\text{fast}} [Co(NH_3)_5Cl]Cl_2 + H_2O$$

In this mechanism, the slow step involves an anion leaving a lattice site and entering the coordination sphere of the metal. Because both the $[Co(NH_3)_5H_2O]^{3+}$ cation and the Cl^- anion are involved in forming the transition state, the process was called S_N2. However, the compound already contains three chloride ions per cation and the formula $[Co(NH_3)_5H_2O]Cl_3$ contains both cation and anions. Therefore, it is not clear what S_N2 means when only the compound $[Co(NH_3)_5H_2O]Cl_3$ is involved.

A mechanism like that postulated above is very unlikely on the basis of energetics. First, an anion must leave an anion site to form a Schottky defect, a high-energy process. Second, the anion must attach to the $[Co(NH_3)_5H_2O]^{3+}$ cation resulting in a 7-bonded transition state. Such complexes require sacrificing a considerable amount of energy in the form of crystal field stabilization energy. As a result, the activation energy would be much larger than the 79 kJ mol^{-1} reported for the deaquation-anation reaction of the chloride compound.

A more realistic approach to the mechanism of this reaction is that H_2O is lost from the coordination sphere of the metal and that it must occupy interstitial positions in the crystal lattice. Diffusion of H_2O from the lattice is favored by the cation and anion having greatly differing sizes because the fraction of free space increases as the difference in size of the cation and anion increases. Thus, the activation energy for loss of H_2O from complexes like $[Cr(NH_5)_5H_2O]Cl_3$ should vary with anion as $Cl^- < Br^- < I^-$, and the reported activation energies are 110.5, 124.3, and 136.8 kJ mol^{-1}, respectively.

Part of the misconception regarding the kinetics of the deaquation-anation reaction stems from the fact that only a limited analysis of the data was performed. To be complete, the data should be analyzed using all of the rate laws shown in Table 7.2. A recent study of this process was completed in which the

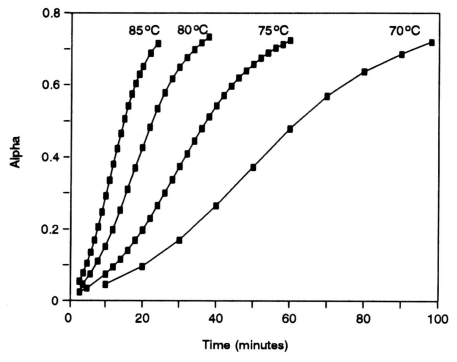

Figure 7.4   Rate plots for the dehydration-anation reaction of $[Co(NH_3)_5H_2O]Cl_3$.

reaction was studied by means of mass loss as the H_2O is lost. Figure 7.4 shows typical rate plots for the process carried out at several temperatures.

The unmistakable sigmoidal nature of the rate plots for deaquation-anation of $[Co(NH_3)_5H_2O]Cl_3$ suggests that the process obeys an Avrami-Erofeev type of rate law. When the data from 32 runs were analyzed by a computer technique that tests all of the rate laws shown in Table 7.2, 26 of the runs gave the best fit with the A1.5 rate law, four gave the best fit with the A2 rate law, and two gave the best fit with the R3 rate law. Not a single run gave the best fit with a first- or second-order rate law. An A1.5 rate law is one form of a nucleation rate law, and it has also been shown to represent a diffusion process. The rate constants obtained using the rate law

$$[-\ln(1-\alpha)]^{2/3} = kt \qquad (7.70)$$

were used to prepare the Arrhenius plot shown in Figure 7.5.

The slope of the line corresponds an activation energy of 97 kJ mol^{-1}, which is somewhat different than the value of 79 kJ mol^{-1} reported by others.

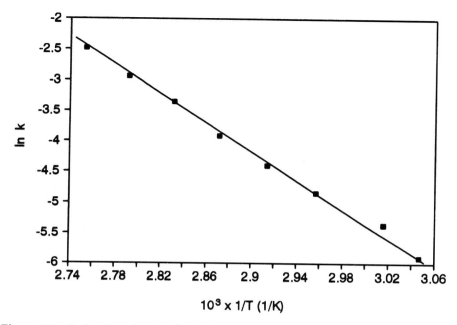

Figure 7.5 Arrhenius plot for the dehydration-anation reaction of $[Co(NH_3)_5H_2O]Cl_3$ when the Avrami-Erofeev rate law with n = 1.5 is used.

This study serves to illustrate how the application of kinetic labels such as S_N1 or S_N2 can be misleading. Only by a careful analysis of data from a large number of runs using a wide range of rate laws can a correct modelling of a reaction be obtained. Unfortunately, this fact has been overlooked many times in the past by workers who have assumed that the kinetics of reactions in the solid state and in solution should be similar.

7.6.2 The Deaquation-Anation Reaction of $[Cr(NH_3)_5H_2O]Br_3$

The deaquation-anation reaction of $[Cr(NH_3)_5H_2O]Br_3$,

$$[Cr(NH_3)_5H_2O]Br_3(s) \longrightarrow [Cr(NH_3)_5Br]Br_2(s) + H_2O(g) \quad \textbf{(7.71)}$$

shows some of the classic problems associated with the study of reactions in the solid state (Ingram, 1995). When the (α,t) data for this reaction were tested using the rate laws shown in Table 7.2, none of the rate laws gave an especially good fit over a wide range of α values. The sigmoidal shape of the (α,t) curves suggested that an Avrami type of rate law should be applicable. In this case, the

(α,t) data for each run were divided into two groups. The first group consisted of data with $0.1 < \alpha < 0.5$, while the second group consisted of data with $0.5 < \alpha < 0.9$. In this way, an attempt was made to identify a rate law that might be applicable to a portion of the reaction.

Four runs were made at each of the temperatures 85, 90, 95, 100, 105, and 110°C. When α was in the range 0.1 to 0.5, 20 out of the 24 runs gave the best fit to the data with an A2: $[-\ln(1 - \alpha)]^{1/2}$ rate law. On the other hand, when α was in the range 0.5 to 0.9, all 24 runs gave the best fit with the D3: $[1-(1 - \alpha)^{1/3}]^2$ rate law. In every case, the correlation coefficient was at least 0.9999. These results suggest that the early stages of the reaction are controlled by the formation of nuclei, which grow in two dimensions, while the latter stages are controlled by the diffusion of water through the lattice. Such results are in accord with defect-diffusion process (see Section 7.6.1). It seems likely in this case that the early loss of water is from near the surface of the particles as the process of nucleation occurs. The latter stages involve the loss of water from the interior of the crystals and thus are controlled by diffusion of the water to the surface. The aspects of this reaction serve to show that even a comparatively simple-looking reaction in the solid state may exhibit peculiarities which make it impossible to predict kinetic behavior.

7.6.3   The Dehydration of *Trans*-[Co(NH$_3$)$_4$Cl$_2$]IO$_3$ · 2H$_2$O

When *trans*-[Co(NH$_3$)$_4$Cl$_2$]IO$_3$ · 2H$_2$O is heated, it loses the water of hydration and isomerizes to *cis*-[Co(NH$_3$)$_4$Cl$_2$]IO$_3$. An early study of this compound indicated that the reaction followed the rate law

$$-\ln(1 - \alpha) = kt + c \qquad (7.72)$$

A later study showed that when the material was heated in thin beds in a nitrogen atmosphere the rate law was

$$-\ln(1 - \alpha) = kt^2 + c \qquad (7.73)$$

and that the reaction followed a first-order rate when the sample bed was made thicker. Also, the activation energy varied from 58 kJ mol^{-1} at a nitrogen pressure of 0.1 torr to 116 kJ mol^{-1} at a nitrogen pressure of 650 torr. In a flowing nitrogen atmosphere, an activation energy of 57 kJ mol^{-1} was found.

A recent study of this reaction was conducted using *trans*-[Co(NH$_3$)$_4$Cl$_2$]IO$_3$· 2H$_2$O having a particle size distribution of 57 ± 15 μm (House and Eveland, 1994). The reaction was studied in the range of temperature from 120 to 140°C, and the (α,t) data were analyzed using the rate laws shown in Table 7.2. Figure 7.6 shows the rate plots obtained when data in the range $0.1 < \alpha < 0.8$ were used.

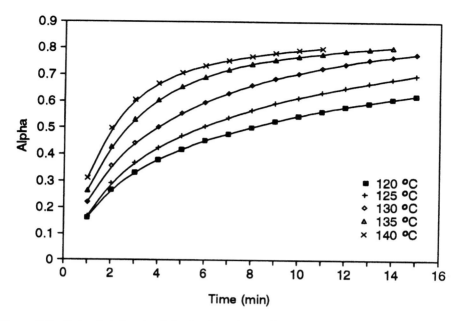

Figure 7.6 Rate plots for the dehydration of *trans*-[Co(NH$_3$)$_4$Cl$_2$]IO$_3$ · 2H$_2$O.

First, it is readily apparent that the rate plots indicate that the process is deceleratory in nature, suggesting an order, diffusion-controlled, or geometric type of rate law. Four runs were made at 120, 125, . . . , 140°C so that a total of 20 runs were analyzed using the rate laws shown in Table 7.2. Twelve of the 20 runs gave the best fit with the second-order rate law

$$1/(1-\alpha) = kt \qquad (7.74)$$

The eight other runs gave the best fit with the third-order rate law,

$$1/(1-\alpha)^2 = kt \qquad (7.75)$$

Using the rate constants obtained from the second-order rate law, an Arrhenius plot gave an activation energy of 103 kJ mol^{-1}, which is very close to the value reported by other workers when the nitrogen pressure is near 1 atm.

In order to investigate the details of this interesting reaction, the early portion (0.1 < α < 0.3) and an intermediate portion (0.3 < α < 0.5) of the rate plots were greatly expanded in order to provide at least 20 data points each for analysis. These smaller regions of the rate plots were analyzed by a computer program testing all of the rate laws shown in Table 7.2. The reaction at 140°C was so rapid that the $0.1 < \alpha < 0.3$ portions could not be analyzed accurately. Therefore, 16 runs were analyzed in that region and 10 of the runs gave the best fit with the third-order (F3) rate law. The other six runs gave the best fit with diffusion-control rate laws, D4(2), D2(3), and D1(1). An F3 rate law probably does not indicate a molecularity of three for the transition state, but it probably indicates that the initial loss of H_2O ($\alpha < 0.3$) is very rapid. An activation energy of 80 kJ mol^{-1} results from this initial loss of water.

In the range $0.3 < \alpha < 0.5$, only two of the 20 runs gave a best fit with the F3 rate law. The remaining 18 runs gave the best fit with D3(7), D2(6), D1(3), and D4(2) rate laws. While there is little doubt that the reaction follows a rate law indicating diffusion control, it is not clear whether it is a D3 or D2 rate law, although there is a slight preference for D3. Using the rate constants from the D3 rate law, an activation energy of 110 kJ mol^{-1} is obtained. Certainly the loss of H_2O from the interior of the crystalline solid could well be expected to be controlled by diffusion.

The fact that the overall process obeys an F2 rate law is probably a compromise between different stages that follow other rate laws even though correlation coefficients in the 0.998 to 0.999 range were found. This study points out that even a material of rather homogeneous particle distribution studied in a highly replicated manner can yield less than complete agreement as to the applicable rate law. It also shows that the study of different regions of the reaction can yield valuable insight as to changes in the mechanism with extent of reaction. As stated earlier (see Section 7.1), it is unlikely that a single rate law will describe the entire reaction, but such is the nature of solid state reactions.

REFERENCES FOR FURTHER READING

Borg, R., Dienes, G. J. (1988) *An Introduction to Solid State Diffusion,* Academic Press, San Diego, Chapters 10 and 12. These chapters deal with rate studies involving defects in solids.

Brown, M. E., Dollimore, D., Galway, A. K., in **Bamford, C. H., and Tipper, D. F.** Eds. (1980) *Comprehensive Chemical Kinetics,* Vol. 22, Elsevier, Amsterdam. An entire volume dealing with all aspects of reactions in solids. Also covers material on techniques of data analysis.

Garner, W. E., Ed. (1955) *Chemistry of the Solid State,* Academic Press, New York. The standard reference on early work on solid state reactions.

Gomes, W. (1961). *Nature (London) 192,* 865. An interesting discussion on the interpretation of the activation energy for a solid state reaction.

Hamilton, D. G., House, J. E. (1994) *Transition Met. Chem. 19,* 527. The kinetic study on the deaquation-anation reaction of aquapentamminecobalt (III) chloride.

House, J. E. (1993) *Coord. Chem. Rev. 128,* 175–191. A review of anation reactions and applications of the defect-diffusion model.

House, J. E. (1980). *Thermochim. Acta 38,* 59. The original description of the defect-diffusion view of how reactions of solid coordination compounds take place.

House, J. E., Eveland, R. W. (1994) *Transition Met. Chem. 19,* 199.

Ingram, B. V. (1995) M. S. Thesis, Illinois State University. A comprehensive study of the deaquation-anation reactions of aquapentamminechromium (III) complexes.

Ng, W.-L., Ho, C.-C., Ng, S.-K. (1978). *J. Inorg. Nucl. Chem. 34,* 459.

O'Brien, P. (1983) *Polyhedron 2,* 233. A review of solid state racemizations and an extension of the ideas presented by the defect-diffusion model.

Schmalzreid, H. (1981) *Solid State Reactions,* 2d ed., Verlag Chemie, Weinheim. A discussion of many principles and types of solid state reactions.

Young, D. A. (1966) *Decomposition of Solids,* Pergamon Press, Oxford. An excellent treatment of decomposition reactions giving derivations of many rate laws.

PROBLEMS

1. Derive the rate law for the reaction of a gas on the surface of solid particles which are cubic with edge length l. Assume that the diffusion of the gas is inversely proportional to the thickness of the product layer. Transform the rate law into one involving α, the extent of reaction. What geometric information does the rate constant contain?

2. Suppose that a gas reacts with solid particles that are thin coin-shaped cylinders, but that the reaction is only on the top and bottom circular faces. Derive the rate law for this reaction assuming that the diffusion of the gas is inversely proportional to the thickness of the product layer. Transform the rate law into one involving α, the extent of reaction. What geometric information does the rate constant contain?

3. In deriving the parabolic rate law, it is assumed that the rate of diffusion of the gaseous reactant is inversely proportional to the thickness of the product layer. Assume that the diffusion of the reactant gas varies as e^{-ax} instead of $1/x$ where a is a constant and x is the thickness of the product layer. Derive the rate law that would result in this case.

4. A solid compound X is transformed into Y when it is heated at 75°C. A sample of X that is quickly heated to 90°C for a very short time (there is no significant decomposition) and then quenched to room temperature is later found to be converted to Y at a rate that is 2.5 times that of an untreated sample of X when both are heated at 75°C for a long period of time. Explain these observations.

5. Suppose that the reaction $A(s) \longrightarrow B(g) + C(s)$ is being studied. Provide a brief discussion of the results and what kind of information the results would provide from studying the effects of each of the following on the rate of the reaction of the $f(\alpha)$ versus time curve.
   (a) Annealing crystals of A
   (b) Running the reaction under high vacuum
   (c) Prior treatment of A with X-rays
   (d) Varying the particle size of A
   (e) Sintering of A as the reaction takes place
6. For the reaction $A(g) + B(s) \longrightarrow C(s)$, the rate is controlled by the diffusion of $A(g)$ through the product layer. Assume that the diffusion is inversely proportional to the thickness of the product layer raised to some power, z. Derive the rate law for this process.
7. For the reaction

$$[Cr(NH_3)_5 H_2O]Cl_3 (s) \longrightarrow [Cr(NH_3)_5 Cl]Cl_2 (s) + H_2O(g)$$

which was carried out at 85°C, the following data were obtained. It is known that this reaction follows an Avrami type of rate law. Test these data to determine n, the index of reaction.

Time (min)	α
0	0
2	0.061
4	0.129
6	0.209
8	0.305
10	0.405
12	0.514
14	0.619
16	0.716
18	0.792
20	0.848
22	0.890
24	0.920
26	0.939
28	0.953

chapter 8

NONISOTHERMAL METHODS IN KINETICS

In recent years, a variety of experimental techniques have been developed which permit studying some changes in a sample as the temperature increases in a linear way with time. As the sample undergoes an increase in temperature, there are numerous changes that take place. For example, the volume almost certainly changes and this provides the technique known as thermal dilatometery. A structural change other than simple thermal expansion of the solid will usually cause an abrupt change in volume. Also, the reflectance spectrum of the solid will likely show evidence of structural changes. In other cases, complexes such as those of Ni^{2+}, a d^8 ion, may undergo a square planar (D_{4h}) to tetrahedral (T_d) isomerization. Because d^8 ions in a square planar ligand environment have no unpaired electron spins and in a tetrahedral environment they have two unpaired electrons, there is a change in magnetic moment of the sample of the complex as its temperature changes in this type of geometrical isomerization.

While these and other specialized methods have become widespread and have seen considerable use, the most widely used methods are thermogravimetric analysis and differential scanning calorimetry. As a group, thermal methods of analysis now constitute the most widely used experimental techniques in the chemical industry. A major reason for this widespread use is that determination of bulk properties, thermal stability, and characterization of materials are as important as the determination of molecular properties.

8.1 TGA AND DSC METHODS

While the properties described above are utilized in specific kinds of experiments, the two most common properties studied are *mass* and *enthalpy*. The change in mass as the temperature increases is produced as a volatile product is released. Therefore, this technique is referred to as *thermogravimetric analysis* (TGA).

In *differential scanning calorimetry* (DSC), the heat flow to the sample is compared to an inert reference as both are heated at the same rate. When an endothermic transition occurs in the sample, the recorder shows a peak, the area of which is proportional to the amount of heat absorbed. When an exothermic transition occurs, the opposite effect is seen. Because the difference in electrical power is monitored as the sample and reference have their temperature changed, what is measured is dH/dT. But $dH/dT = C_p$, the heat capacity at constant pressure and

$$\Delta H = \int C_p dT \qquad (8.1)$$

Therefore, the peak area gives directly the value of ΔH after a calibration in terms of cal/in^2 or J/in^2 is made using a reaction of known enthalpy. Any type of change that absorbs or liberates heat can be studied using DSC (fusion, phase transition, decomposition, etc.) while in TGA experiments only those changes that involve a mass loss can be studied.

Although several thermoanalytical methods are of potential use in studying reactions occurring in the solid state, TGA and DSC are most often used. Sophisticated equipment, complete with interfaced microcomputer, is available and these methods are finding wide use in the study of solid materials. Because both of these methods can readily yield data which are suitable for kinetic analysis, their use for that purpose will be described in greater detail. It should be pointed out that this is an area of intense research activity, and the literature in this field is growing at a very rapid rate. The methods that will be described briefly here are included only to show representative examples of the techniques used. Many others have been published and an entire book could easily be written describing nonisothermal kinetic methods and their areas of application. Thermal analyses are especially invaluable in the polymer industry.

As mentioned above, a TGA experiment determines the mass of the sample, either as its temperature is held constant or as it is changed in some programmed way. The mass is measured using a microbalance, which is capable of determining a mass loss of 10^{-6} grams. TGA is a most useful technique when a reaction of the type

$$A(s) \longrightarrow B(s) + C(g) \qquad (8.2)$$

is being studied. In this kind of process, which is characteristic of numerous solid state reactions, the extent of mass loss can be used to establish the stoichiometry of the reaction. Consider a complex represented as $[M(NH_3)_6]X_2$ where M is a metal ion and X is an anion such as Cl^-. Heating a complex of this type in a TGA could produce a curve such as that shown in Figure 8.1.

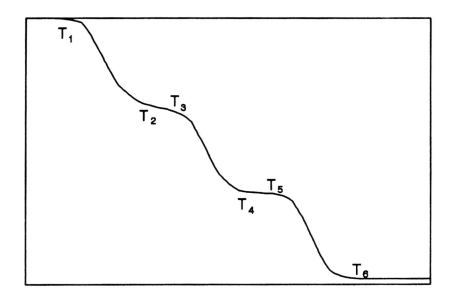

Figure 8.1   A hypothetical TGA curve for loss of six ammonia molecules in three steps from a hexammine complex.

When complexes such as $[M(NH_3)_6]X_2$ are heated, there is usually a stepwise loss of the volatile ligands, NH_3. The three mass plateaus shown in Figure 8.1 are indicative of the reactions. With the initial and final temperatures being indicated, these reactions could be

$$[M(NH_3)_6]X_2(s) \xrightarrow{\ T_1 - T_2\ } M(NH_3)_4 X_2(s) + 2\,NH_3(g) \tag{8.3}$$

$$M(NH_3)_4 X_2(s) \xrightarrow{\ T_3 - T_4\ } M(NH_3)_2 X_2(s) + 2\,NH_3(g) \tag{8.4}$$

$$M(NH_3)_2 X_2(s) \xrightarrow{\ T_5 - T_6\ } MX_2(s) + 2\,NH_3(s) \tag{8.5}$$

Whether these are, in fact, the actual reactions must be determined by comparing the observed mass losses with those expected for the reactions.

Frequently, it is not easy to determine the temperatures $T_1, T_2, \ldots$, owing to the nearly horizontal nature of the curve in those regions. This means also that it is not easy to determine accurately the mass loss in such cases. Because of this, some instruments also compute the trace of the first derivative of the mass curve. This DTG (derivative thermogravimetric) curve makes it much easier to determine where a zero slope indicates that the sample is undergoing no change in mass. Therefore, the reaction stoichiometry can be more accurately determined.

8.2   KINETIC ANALYSIS BY THE COATS AND REDFERN METHOD

While reaction stoichiometry is an important use of TGA, our purpose here is to investigate its use in studying reaction kinetics. This is done by employing a rate law which is usually of the form

$$\frac{d\alpha}{dt} = k(1-\alpha)^n \tag{8.6}$$

where α is the fraction of the reaction completed, t is the time, n is the "reaction order" (more properly, an "index of reaction"), and k is the rate constant. The fraction of the reaction completed is computed by dividing the actual mass loss at some temperature by the mass loss expected for a complete reaction. Thus, if a certain reaction corresponds to a mass loss of 40% and the observed mass loss at some temperature is 10%, the value of α at that temperature is $10/40 = 0.25$.

   In DSC, the extent of the reaction is obtained by first determining the total peak area corresponding to the complete reaction. At a specific temperature, the partial peak area is determined and the fraction of the reaction complete at that temperature is determined by comparing the partial area to the total peak area. Again, the object of the experiment is to be able to determine the fraction of the reaction complete as a function of temperature so that nonisothermal kinetics procedures can be applied.

   The rate constant for most chemical reactions can be expressed by the Arrhenius equation,

$$k = A\ e^{-E/RT} \tag{8.7}$$

where E is the activation energy, A is the frequency factor (assumed here to be independent of temperature, which is not always the case), and R is the molar gas constant. If the heating rate (β) is constant, β (deg/min) = dT/dt, which can be written as dt = dT/β.

   After substituting for dt and k in the rate equation, rearrangement leads to

$$\frac{d\alpha}{(1-\alpha)^n} = \frac{A}{\beta} e^{-E/RT} dT \tag{8.8}$$

Then,

$$\int_0^\alpha \frac{d\alpha}{(1-\alpha)^n} = \frac{A}{\beta} \int_0^T e^{-E/RT} dT \tag{8.9}$$

The right-hand side of this equation can not be integrated directly because the integral known as the temperature integral,

$$I = \int_0^T e^{-E/RT} dT \tag{8.10}$$

has no exact analytical form. The left-hand side can be integrated for various values of n. It is the multitude of ways of approximating the temperature integral that gives rise to a large number of ways of analyzing TGA data.

One of the most common ways of handling this problem is that of approximating the temperature integral as a series and then truncating it after a small number of terms. When this is done, the result after taking logarithms is,

$$\ln \frac{1-(1-\alpha)^{1-n}}{(1-n)T^2} = \ln \frac{AR}{\beta E}\left(1 - \frac{2RT}{E}\right) - \frac{E}{RT} \tag{8.11}$$

when n is not equal to one, or, when $n = 1$,

$$\ln \ln \frac{1}{1-\alpha} - 2 \ln T = \ln \frac{AR}{\beta E}\left(1 - \frac{2RT}{E}\right) - \frac{E}{RT} \tag{8.12}$$

At room temperature, 2RT is about 5 kJ mol^{-1} (1.2 kcal mol^{-1}) so that for most reactions E >> 2RT, and the term

$$\ln \frac{AR}{\beta E}\left(1 - \frac{2RT}{E}\right)$$

is treated as a constant. Therefore, the rate equations have the form of a linear relationship when the left-hand side is plotted against $1/T$. The intercept is $\ln[(AR/\beta E)(1 - (2RT/E))]$, from which the frequency factor (A) can be computed, and the slope is $-E/R$.

The way one proceeds in practice is to determine a series of α_i values at a series of temperatures, T_i. Then, the function represented by the left-hand side of the equation is computed for each pair of (α,T) values for various values of n. The $f(\alpha,T)$ values are plotted against $1/T$, and the series that gives the most linear plot is presumed to correspond to the "correct" reaction index, n. Table 8.1 shows (α, T) data for a hypothetical reaction. Eqs. (8.11) and (8.12) were used to calculate the $f(\alpha,T)$ values to plot against $1/T$ when different trial n values were used (0, 1/3, 2/3, 1, and 2 in this case) and these values are also shown in Table

Table 8.1   Values of α and f(α) for various trial *n* values.

T(K)	$10^3/T$	Alpha	$n = 0$	$n = 1/3$	$n = 2/3$	$n = 1$	$n = 2$
				f(α)			
410	2.439	0.06974	−14.695	−14.683	−14.671	−14.659	−14.623
420	2.381	0.14317	−14.024	−13.999	−13.974	−13.948	−13.870
430	2.326	0.27660	−13.413	−13.361	−13.309	−13.255	−13.089
440	2.273	0.49189	−12.883	−12.781	−12.674	−12.564	−12.206
450	2.222	0.77010	−12.480	−12.283	−12.068	−11.833	−11.010
460	2.174	0.97868	−12.284	−11.937	−11.489	−10.915	−8.436

8.1. The best fit is provided when $n = 2/3$, so the reaction is called 2/3-"order." This procedure is illustrated in Figure 8.2.

While this procedure may certainly give an "optimum" value of *n* with respect to the linearity of the f(α,T) versus $1/T$ plot, the "order" may have no actual relationship to the molecularity of a transition state in the usual kinetic sense. The *n* value is usually called the index of reaction. In most cases, the results obtained from this type of analysis are similar to the kinetic parameters obtained by conventional isothermal means, and in many cases the agreement is excellent. However, it must be remembered that the original rate law assumed the form

$$\frac{d\alpha}{dt} = k(1 - \alpha)^n$$

as in Eq. (8.6) and, as we have seen, there are approximately 20 possible rate laws for solid state reactions, many of which can not be put in this form (see Table 7.2). Therefore, it may be that a good fit to the data can be obtained for some value of *n* although the actual rate law should be of some other form. This will be discussed more fully later.

The nonisothermal kinetic method described above is known as the Coats and Redfern method (Coats and Redfern, 1964), but it is by no means the only such method. It is one of the most straightforward methods, and the entire process can be programmed for use on a microcomputer or programmable calculator. In this case, the (α,T), data are entered only once and the entire sequence of *n* values is tested automatically. The value of *n* giving the best fit is identified by correlation coefficient from the linear regression procedure. Theoretical values of *n* are 0, 1/3, 1/2, 2/3, 1, and 2 because these are justifiable on the basis of geometrical, contracting, or reaction order models (see Chapter 7). Also, it is possible to treat *n* as simply an exponent to be varied, and iterative

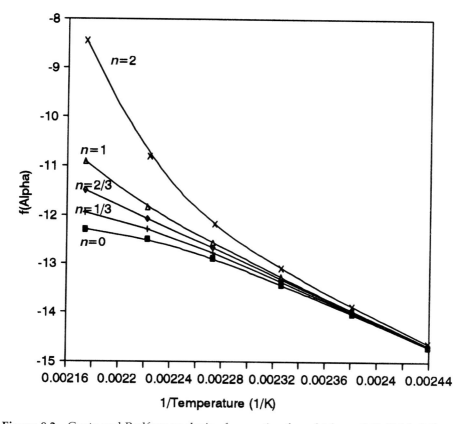

Figure 8.2   Coats and Redfern analysis of a reaction for which $n = 2/3$ (Table 8.1).

procedures have been devised to identify any value of n that provides the best fit to the data. It is by no means clear what an n value of 0.43 or 1.12 means in terms of mechanism, however.

8.3   THE REICH AND STIVALA METHOD

An interesting variation of the Coats and Redfern method has been developed by Reich and Stivala that makes use of an iterative technique to arrive at the best value of n. It is best employed using a computer to perform all of the computations. The integrated rate equation is written in the form

$$\frac{1 - (1-\alpha)^{1-n}}{1-n} = \frac{ART^2}{\beta E}\left(1 - \frac{2RT}{E}\right)e^{-E/RT}$$

(8.13)

where the temperature integral has been written as a truncated series as before. Again we recognize that E >> 2RT so that [1 – (2RT/E)] is very nearly a constant. Therefore, the equation can be written in the two-point form

$$\ln\left[\frac{1-(1-\alpha_i)^{1-n}}{1-(1-\alpha_{i+1})^{1-n}}\left(\frac{T_{i+1}}{T_i}\right)^2\right] = -\frac{E}{R}\left(\frac{1}{T_i} - \frac{1}{T_{i+1}}\right) \tag{8.14}$$

This is the form of a linear equation when we let y represent $f(\alpha,T)$ on the left-hand side and x represent $[(1/T_i) - (1/T_{i+1})]$. The slope will be $-E/R$, and the intercept will be zero. However, this condition will be met only when the correct value of n is used. For N pairs of (α,T) data, there will be (N – 1) values of x and y to compute. Linear regression is performed on these (x,y) data pairs for an initial trial value of n. The form of the equation is such that the intercept will be zero for the correct value of n, but it has a positive and decreasing value as n approaches the correct value. An initial value of $n = 0.1$ is chosen and the computations performed with the intercept being compared to zero. If the intercept is greater than zero, n is incremented by 0.1, and the process is repeated. When a value of n results in a *negative* intercept, this means that the "correct" value of n is between that value and the previous value. Therefore, n is decremented to its previous value and step refinement is accomplished by making the increment in n equal to 0.01. The iterative procedure is again implemented until the intercept undergoes another sign change, indicating that the "correct" value of n has been passed. At this point, the computation is either repeated using a smaller (0.001) increment to n, or the process is terminated. For all real data, there is no need to try to obtain an n value to three decimal places. It is certainly not justifiable to try to attach any "order" significance to a value of say $n = 0.837$!

8.4   A METHOD BASED ON THREE (α,T) DATA PAIRS

Another method to obtain n is possible that is based on the same two-point form of the Coats and Redfern equation, which was also used by Reich and Stivala (Eq. 8.14). It is easy to see from inspection of Eq. (8.14) that by using pairs of (α,T) data, a constant value of E/R will be obtained only if n has the correct value. This is illustrated by the data shown in Table 8.2 which presents calculated values of α obtained by numerical solution of Eq. (8.8) for specific values of n, E, and A. The calculated E/R values shown are based on various assumed values of n. The calculated values of α are much more accurate than any that can be obtained experimentally.

From the data shown in Table 8.2, it is clear that the value of E/R is constant only for the correct value of n (in this case 5/3, or 1.667). It is also clear

Table 8.2 Values of (α, T) and E/R computed using $n = 5/3$, $E = 100$ kJ/mol and $A/\beta = 3 \times 10^{10}$ min^{-1}. A fourth-order Runge-Kutta method was used to compute α from the differential rate equation.

					E/R values			
Point	T_1 (K)	α_1	T_2(K)	α_2	$n = 0.4$	$n = 1.2$	$n = 5/3$	$n = 2$
1	400	0.03177	410	0.06740	11646	11891	12036	12140
2	410	0.06740	420	0.13369	11210	11715	12016	12233
3	420	0.13369	430	0.24365	10452	11419	12009	12441
4	430	0.24365	440	0.39894	9238	10931	12003	12807
5	440	0.39894	450	0.57706	7559	10216	12000	13378
6	450	0.57760	460	0.73851	5619	9311	11999	14150
7	460	0.73851	470	0.85531	3751	8321	11997	15039
8	470	0.85531	480	0.92584	2220	7371	11995	15916
9	480	0.92584	490	0.96354	1104	6539	11991	16668
10	490	0.96354	500	0.98235	347	5857	11994	17258

Note: For $E = 100$ kJ/mol and $R = 8.31443$ J/mol K, $E/R = 12027$ K^{-1}.

that for small values of α, the value of E/R does not change much regardless of the value of n chosen. The first step in the analysis is to compute an approximate value of E/R, and this is performed using two data pairs, (α_1, T_1) and (α_2, T_2), where α_1 and α_2 are small. The value of E/R is nearly constant for any value of n under these conditions. The first value assigned to the index of reaction, n_o, is zero, and a reasonably good value of E/R results with n_o as long as α_1 and α_2 are small (see rows 1 and 2 of E/R values shown in Table 8.2). For the first two values, the E/R value varies only from 11,646 to 12,140 as n varies from 0.4 to 2.

After getting an approximate first iterate value for E/R, two data points (α_2, T_2) and (α_3, T_3) are considered where $\alpha_3 \gg \alpha_2$. If $\alpha_3 \gg \alpha_2$, maximum variation of the function occurs as n is iterated. Therefore, a recalculation of E/R using these data when one α value is large will give a value close to the test E/R computed with the first two points only when n has approximately the correct value. For example, using $\alpha_2 = 0.06740$ and $\alpha_3 = 0.98235$, E/R varies from 347 for $n = 0.4$ to 17,258 for $n = 2$. Clearly, because the correct value of E/R is about 12,000, this value will be obtained only when $n \approx 1.67$, and the calculated E/R value is very sensitive to the n value. Using the first iterate E/R value, the function

$$F_1 = \exp\left[\frac{E}{R}\left(\frac{1}{T_3} - \frac{1}{T_2}\right)\right] \Big/ \frac{T_3}{T_2} \qquad \textbf{(8.15)}$$

is calculated using T_2 and T_3. Then, the function

$$F_2 = \frac{[1-(1-\alpha_2)^{1-n}]}{[1-(1-\alpha_3)^{1-n}]} \tag{8.16}$$

is computed with $n_o = 0$ and the result is compared to the value of F_1. It is easily shown that if $F_2 > F_1$, then the iterated reaction index, n', is smaller than the "correct" n ($n' < n$ when $F_2 > F_1$). The process continues by incrementing n' by 0.100001 (so that n' is never exactly equal to 1) and repeating the calculations. At the point where $F_1 > F_2$, n' is greater than n by an amount less than 0.1 for the first iterate value of E/R. This fixes an approximate upper limit of n', usually within 0.1 of the "correct" n. At this point, the value of n' calculated from (α_2, T_2) and (α_3, T_3) using the first iterate E/R value is reduced by 0.1 and the increment to n' is reduced from 0.1 to 0.01 as a step refinement. The resulting n' is used to recalculate a second iterate E/R value using the first two data pairs. This E/R value is very nearly the correct one because the value of n' used is correct to within approximately 0.1.

Having a nearly exact second iterate to E/R, the function F_1 is calculated using points (α_2, T_2) and (α_3, T_3). Processing continues by computing the function F_2 iterating with $\Delta n = 0.01$. When $F_2 < F_1$ occurs, computation ends and n' has been determined with an upper limit within 0.01 of the "correct" value. If desired, the entire process can be repeated to obtain a third iterate to E/R using the last value of n', which is very close to the correct value, and an increment of $\Delta n = 0.001$. This is never really necessary, however. The method is very compact and requires a minimum of storage space in the computing machine. Therefore, it is adaptable to programmable calculators having small capacity. Exhaustive calculations using simulated errors have shown that this method actually yields values of n and E which are less sensitive to errors in α than are those from the Coats and Redfern or Reich and Stivala methods.

8.5 A METHOD BASED ON FOUR (α,T) DATA PAIRS

Another method of determining n and E from (α,T) data makes use of four data pairs. In this method, it is assumed that four values, α_1, α_2, α_3, and α_4, are known at four temperatures, T_1, T_2, T_3, and T_4, respectively. Further, we will assume that $\alpha_1 < \alpha_2 < \alpha_3 < \alpha_4$. The method makes use of the two-point form of the Coats and Redfern equation written as

$$\frac{\ln\left[\frac{[1-(1-\alpha_1)^{1-n}]}{[1-(1-\alpha_2)^{1-n}]}\left(\frac{T_2}{T_1}\right)^2\right]}{\frac{1}{T_2}-\frac{1}{T_1}} = \frac{E}{R} \tag{8.17}$$

By considering an analogous equation for the points (α_3, T_3) and (α_4, T_4), elimination of E/R yields

$$\frac{\ln\left[\frac{[1-(1-\alpha_3)^{1-n}]}{[1-(1-\alpha_4)^{1-n}]}\left(\frac{T_4}{T_3}\right)^2\right]}{\frac{1}{T_4}-\frac{1}{T_3}} = \frac{\ln\left[\frac{[1-(1-\alpha_1)^{1-n}]}{[1-(1-\alpha_2)^{1-n}]}\left(\frac{T_2}{T_1}\right)^2\right]}{\frac{1}{T_2}-\frac{1}{T_1}} \tag{8.18}$$

However, the two sides of this equation will be equal only when n has the correct value. Therefore, it is a simple matter to set up a calculation in which the value of n is iterated until the two sides are equal using the same type of iteration on n described above. A graphical method using this approach has also been described in which the left-hand and right-hand sides of the equation are plotted as functions of n. The point of intersection of the two curves yields the correct value of n. The advantages of this method are similar to those described for the three-point method. It is a compact method that can easily be programmed on a machine of limited capacity. Also, it is more resistant to errors in α than is either the Coats and Redfern or Reich and Stivala methods which are both based on the same general rate law shown in Eq. (8.6).

8.6 A DIFFERENTIAL METHOD

Some of the early methods of data analysis to extract kinetic parameters were based on differential methods of analysis. This refers to the fact that no attempt is made to integrate the rate law, but rather a differential form is used directly. If a reaction follows the rate law

$$\frac{d\alpha}{dt} = Ae^{-E/RT}(1-\alpha)^n \tag{8.19}$$

and the heating rate, β, is dT/dt, this rate law for the reaction studied non-isothermally becomes

$$\frac{d\alpha}{dT} = \frac{A}{\beta}e^{-E/RT}(1-\alpha)^n \tag{8.20}$$

Taking the logarithm of both sides of the rate law gives

$$\ln\frac{d\alpha}{dT} = \ln\frac{A}{\beta} - \frac{E}{RT} + n\ln(1-\alpha) \tag{8.21}$$

Written in terms of the derivatives, this equation becomes

$$d \ln \frac{d\alpha}{dT} = 0 - \frac{E}{R} d\left(\frac{1}{T}\right) + n \, d \ln(1-\alpha) \tag{8.22}$$

Therefore,

$$\frac{d\left(\ln \dfrac{d\alpha}{dT}\right)}{d \ln(1-\alpha)} = -\frac{E}{R} \frac{d\left(\dfrac{1}{T}\right)}{d \ln(1-\alpha)} + n \tag{8.23}$$

Processing the data by plotting d ln (dα/dT)/d ln(1 – α) versus d(1/T)/d ln(1 – α) gives a linear graph having a slope of –E/R and an intercept of n.

8.7   A COMPREHENSIVE NONISOTHERMAL KINETIC METHOD

A very large number of other methods have been developed for treating (α,T) data from nonisothermal kinetic studies to yield kinetic information. Many of them are based on the rate law shown in Eq. (8.6), which is an incomplete equation because it can not be put in a form to describe diffusion control or Avrami type of processes (see Chapter 7). In 1983, Reich and Stivala removed the constraint imposed by Eq. (8.6) by developing a kinetic analysis method that tests most of the common types of rate laws including Avrami, diffusion control and others not covered by Eq. (8.6). The method is based on a computer program which fits the (α,T) data to the rate laws and computes the standard error of estimate (SEE) for each so that the rate law providing the best fit to the data can be identified. It is still true that when a large number of runs are considered, it is rare that a given rate law fits the data for all the runs. It is still necessary to make a large number of runs and examine the results to determine the rate law that fits the data from most of the runs. The 1983 method of Reich and Stivala is still one of the most powerful and useful techniques for studying kinetics of solid state reactions using nonisothermal techniques. The article describing the method also has a listing of the computer program for implementing the procedure (Reich and Stivala, 1983). In general, this method gives excellent agreement with the results of studies where data from isothermal experiments are fitted to the rate laws shown in Table 7.2. In addition to the methods discussed here and in Section 8.3, Reich and Stivala have described a rather large number of data analysis procedures for use in analyzing data from nonisothermal techniques.

8.8   THE GENERAL RATE LAW AND COMPREHENSIVE METHOD

Examination of the rate laws shown in Table 7.2 shows that $(1 - \alpha)$, $-\ln(1 - \alpha)$, and α occur in various combinations and with various exponents. Therefore, a general rate law which describes almost any reaction taking place in the solid state can be written as

$$\frac{d\alpha}{dT} = \frac{A}{\beta} \alpha^m (1-\alpha)^n [-\ln(1-\alpha)]^p e^{-E/RT} \tag{8.24}$$

where β is the heating rate and the other symbols have their usual meanings. If one takes the natural logarithm of both sides of the equation, the result is

$$\ln \frac{d\alpha}{dT} = \ln \frac{A}{\beta} + m \ln \alpha + n \ln (1-\alpha) + p \ln [-\ln(1-\alpha)] - \frac{E}{RT} \tag{8.25}$$

In this equation, m, n, and p as well as E and A are unknowns. This suggests that if α and $d\alpha/dT$ were known at five temperatures, a system of five equations in five unknowns could be solved to find m, n, p, A, and E. Of course it would be preferable to have an over-determined system by having many α and $(d\alpha/dT)$ data points. Such a procedure would enable one to identify any of the rate laws shown in Table 7.2 by determining the appropriate exponents, m, n, and p. If m and p are both zero and $n = 2$, a second order process is indicated. If m and n are both zero and $p = 1/2$, an A2 Avrami rate law is indicated.

In Chapter 7, it was shown that in the analysis of (α,t) data to determine the best-fitting model law that the *same* rate law was not necessarily indicated by *all* runs and that numerous runs might be required. An elaborate computer program has been developed to solve the system of equations using Gauss-Jordan condensation with pivotal rotation (Lowery, 1986). To test the procedure, test data were generated by solving Eq. (8.24) numerically using a fourth order Runge-Kutta technique. In performing the calculations, solutions were carried out with various combinations of exponents, m, n, and p (e.g., $m = 0$, $n = 1/3$, $p = 0$; $m = 1/3$, $n = 0$, $p = 0$; $m = 0$, $n = 1/3$, $p = 1/3$, etc.). This was done so that the (α,T) data from a variety of rate laws could be tested. Also, the general analysis procedure was designed so that the test data could be analyzed not only with a procedure that would determine three exponents, but also one which would keep some of the exponents at zero and "force" the fitting of the data to special case rate laws in which only the other exponents could vary. For example, if $m = p = 0$ and only n can vary, the data are forced to fit a rate law of the form

$$\frac{d\alpha}{dT} = \frac{A}{\beta} (1-\alpha)^n e^{-E/RT}$$

If $m = n = 0$ and only p can vary, the data are forced to fit a rate law of the form

$$\frac{d\alpha}{dT} = \frac{A}{\beta}[-\ln(1-\alpha)]^p e^{-E/RT}$$

which is the form of the Avrami rate laws. Obviously, if the (α,T) data were calculated using $n = 1/3$ and $m = p = 0$, then the (α,T) data were analyzed using a method that forced them to fit a rate law in which only p could vary, some "interesting" results should be obtained. In order to provide a comparison with other methods, the (α,T) data were also analyzed using the Coats and Redfern method (Eq. (8.11)) and the Reich and Stivala (Eq. (8.14)) method based on the same rate law.

How well the procedure works is illustrated by the following cases. Table 8.3 shows the results obtained by applying the general procedure to (α,T) data which were derived from a rate law that is of the form involving a single exponent $m = 0.333$. In calculating the values used in this case, $n = p = 0$ was assumed. The results shown in the table clearly indicate that any time the general procedure allowed for m to be one of the exponents to be determined, the fit is very good. However, if the procedure used was one in which m was omitted from the computation (n, p, or np type of rate law), the fit was quite poor, as expected. Finally, when the procedure used a rate law of the mnp type (all three exponents being determined), the fit was acceptable. In the numerical solution of the differential equation, $E = 100$ kJ mol^{-1} was used and that value was successfully determined by any of the rate laws that tested an "m-type" rate law. Similar results are shown in Tables 8.4 and 8.5 for rate laws of the "n-type" and the "p-type." In each case, the general procedure in which m, n, and p were determined or any abbreviated procedure that contained the appropriate rate law exponent successfully analyzed the data to reproduce the values of E, A, and exponents.

Table 8.3 Results obtained from the analysis of data derived using an α^m type of rate law ($m = 0.333$).

Type[a]	E(kJ mol^{-1})	m	n	p	S[b]
m	100.2	0.332	0.000	0.000	0.36×10^{-9}
n	152.9	0.000	−0.003	0.000	0.48×10^{-6}
p	151.7	0.000	0.000	0.008	0.42×10^{-6}
mn	100.1	0.333	0.000	0.000	0.36×10^{-9}
np	136.5	0.000	0.042	0.102	0.24×10^{-7}
mnp	97.9	0.354	−0.003	−0.007	0.24×10^{-9}

[a]Type refers to the exponents allowed to vary in the data analysis procedure used to analyze the data.

[b]S is the sum of squares of errors from regression fitting of the rate laws.

Table 8.4 Results obtained using the general analysis of data derived using an $(1 - \alpha)^n$ type of rate law ($n = 0.333$).

Type[a]	E(kJ mol^{-1})	m	n	p	S[b]
m	−237.1	3.186	0.000	0.000	0.26×10^{-2}
n	100.0	0.000	0.333	0.000	0.84×10^{-9}
p	227.5	0.000	0.000	−1.18	0.26×10^{-3}
mn	100.0	0.000	0.333	0.000	0.84×10^{-9}
np	100.0	0.000	0.333	0.000	0.84×10^{-1}
mnp	101.2	−0.005	0.332	−0.006	0.77×10^{-9}

[a]Type refers to the exponents allowed to vary in the data analysis procedure used to analyze the data.

[b]S is the sum of squares of errors from regression fitting of the rate laws.

Table 8.5 Results obtained by analysis of data derived using an $[-\ln(1 - \alpha)]^p$ type of rate law ($p = 0.333$ (Avrami, A3)).

Type[a]	E(kJ mol^{-1})	m	n	p	S[b]
m	−187.7	3.219	0.000	0.000	0.11×10^{-3}
n	73.8	0.000	0.121	0.000	0.51×10^{-4}
p	100.0	0.000	0.000	0.333	0.46×10^{-8}
mn	31.1	1.311	0.072	0.000	0.25×10^{-5}
np	99.9	0.000	0.000	0.333	0.44×10^{-8}
mnp	97.9	0.021	0.001	0.328	0.38×10^{-8}

[a]Type refers to the exponents allowed to vary in the data analysis procedure used to analyze the data.

[b]S is the sum of squares of errors from regression fitting of the rate laws.

For the same calculated α, T, and $d\alpha/dT$ data that gave the results shown in Table 8.3 when analyzed by the comprehensive method, the Coats and Redfern method gave an n value of approximately zero, an activation energy of 152 kJ mol^{-1}, and a correlation coefficient of 1.000! The reason for these totally fictitious results is that the rate law

$$\frac{d\alpha}{dT} = \frac{A}{\beta}(1 - \alpha)^n e^{-E/RT}$$

can not be put in a form that represents a rate law based on α^m, the actual rate law used to originate the α and $d\alpha/dT$ data. However, the data *may* be fit by the Coats and Redfern equation with *some* value of n.

Table 8.4 shows the results obtained when the rate law used to determine the (α,T) data was of the "n-type" and those data were analyzed using the complete procedure and those that involve incomplete forms. To obtain the α, T, and $d\alpha/dT$ data that were analyzed to give the results shown in Table 8.4, the rate law

$$\frac{d\alpha}{dT} = \frac{A}{\beta}(1-\alpha)^{1/3}\,e^{-E/RT}$$

was solved numerically. This rate law is of exactly the same form as that used in the Coats and Redfern method, so that method of analysis would be expected to return the input values for n and E. The actual results were $n = 0.33$ and $E = 99.7$ kJ/mol. In this case, either the robust calculation to determine m, n, and p or the method of Coats and Redfern will yield equally reliable results.

In Table 8.5, the results obtained using α, T, and $d\alpha/dT$ data that fit an Avrami type of rate law are shown. The general method can, of course, be used to fit rate laws containing any combination of m, n, and p exponents. Clearly, any of the procedures testing a rate law containing p work well. When analysis of the same (α,T) data by the method of Coats and Redfern is attempted, the results are $n = 0.125$ and $E = 72.9$ kJ/mol. There is no agreement between these values and the input data because the Coats and Redfern method is based on an equation that can not represent a rate law of the A3: $[-\ln(1 - \alpha)]^{1/3}$ type, which was used to generate the test data.

When methods such as the Coats and Redfern method and others that are based on the same rate law are used to analyze (α,T) data, adequate results are obtained *if* the rate law being followed is one that can be represented by

$$\frac{d\alpha}{dT} = \frac{A}{\beta}(1-\alpha)^{n}e^{-E/RT}$$

However, if the (α,T) data are from a process that follows some other rate law (Avrami, diffusion control, etc.), application of the Coats and Redfern and all similar methods will give erroneous results *even if the correlation coefficient is 1.000*. In the past, many studies have not taken this into account, and it has been assumed that a good fit by Coats and Redfern plots assures that a correct rate law has been identified when in fact the actual rate law may be different. While calculated data based on numerous other combinations of exponents were analyzed, the results above serve to show the application of the method based on the comprehensive rate law. The results obtained when two of the three exponents were used were similar. For example, in one case where the exponents used to determine the $(\alpha, d\alpha/dT, T)$ data were $m = 0.333$, $n = 0.333$, and $p = 0.333$, the robust calculation returned the values 0.334, 0.334, and 0.331, respectively, and an activation energy of 99.6 kJ mol^{-1}. Obviously, the complete procedure can determine the exponents for almost any rate law.

This situation does not necessarily mean that *all* kinetic data which have been obtained by Coats and Redfern and similar methods are incorrect. For example, the calculated activation energy frequently has about the same value regardless of whether the correct rate law has been identified or not. That is because the rate of the reaction responds to a change in temperature according to the Arrhenius equation. The rate law used to fit the kinetic data does not alter the influence of temperature. Also, many kinetic studies on reactions in the solid state have dealt with series of reactions using similar compounds. As long as a consistent kinetic analysis procedure is used, the trends within the series will usually be valid. Undoubtedly, however, many studies based on incomplete data analysis procedures have yielded incorrect kinetic parameters and certainly have yielded no information on mechanism.

The results described in this section show that the general method is quite successful at identifying the correct exponents and activation energy when a rate law that contains the correct exponent(s) is used. There is, however, a serious problem. The amount of mathematical apparatus is such that very accurate values of α and $d\alpha/dT$ are needed to get reasonably accurate values of the exponents m, n, and p. Without the experimental input data being sufficiently accurate, it is still not possible to uniquely identify the correct rate law. There are sufficient experimental errors and sample-to-sample variations that it is still not possible to identify uniquely a rate law using experimental data even when state-of-the-art equipment is used. The computer procedure can determine a set of exponents that provide a reasonably good fit to the data, but these constants will likely have enough uncertainty that the rate law will not be known. Although no evidence will be presented here, the calculations seem to be most sensitive to the values of $d\alpha/dT$. The techniques used in these studies, especially TGA and DSC, have improved enormously in the last decade but these problems have not been completely eliminated. Clearly, while great strides have been made in the treatment of data from nonisothermal experiments and kinetic studies on reactions in the solid state yield much information, this branch of chemical kinetics still needs additional development before it will become an exact science. In spite of the difficulties, the best nonisothermal kinetic studies yield results that are comparable in quality to isothermal kinetic studies on solid state reactions.

REFERENCES FOR FURTHER READING

Brown, M. E. (1988) *Introduction to Thermal Analysis,* Chapman and Hall, London. A text describing the several types of thermal analysis and their areas of application. Chapter 13 is devoted to nonisothermal kinetics.

Brown, M. E., Phillpotts, C. A. R. (1978) *J. Chem. Educ. 55,* 556. An introduction to nonisothermal kinetics.

Coats, A. W., Redfern, J. P. (1964) *Nature (London), 201,* 68. The original description of one of the most widely used methods for analysis of nonisothermal kinetics.

Lowery, M. D. (1986) *M. S. Thesis,* Illinois State University. Complete details of a comprehensive analysis procedure based on the general kinetic equation.

House, J. E., House, J. D. (1983) *Thermochim. Acta 61,* 277. A description of the three-point method and its applications.

House, J. E., Tcheng, D. K. (1983) *Thermochim. Acta, 64,* 195. A description of the four-point method and its applications.

Reich, L., Stivala, S. S. (1980) *Thermochim. Acta, 36,* 103. The computer method of applying the Coats and Redfern method iteratively.

Reich, L., Stivala, S. S. (1983) *Thermochim Acta, 62,* 129. The description of a versatile nonisothermal kinetic method complete with a program listing in BASIC.

Wunderlich, B. (1990) *Thermal Analysis,* Academic Press, San Diego. An introduction to thermal methods of analysis and their uses in numerous areas of chemistry.

PROBLEMS

1. For a certain reaction, the following data were obtained.

T, K	383	393	403	407	413	417	423
α	0.107	0.208	0.343	0.410	0.535	0.623	0.765

   Analyze these data using the Coats and Redfern method to determine n and E.

2. Analyze by the Coats and Redfern method the following data for a solid state reaction to determine n, E, and A.

T, K	390	400	410	420	430	440	450
α	0.014	0.032	0.070	0.145	0.283	0.514	0.824

3. Assuming a rate law of the form $d\alpha/dt = k(1 - \alpha)^n$, obtain the integrated rate laws for the values $n = 0, 1/2, 1$, and 2.

4. Use the four data points given to determine the approximate value of n for the reaction A $\longrightarrow$ B. Try a few values of n (0, 2/3, 4/3, and 2 should be adequate) with the appropriate functions and make plots to determine the intersection point.

T, K	390	410	430	450
α	0.0143	0.0702	0.2834	0.845

   Having determined n, describe the type of process that is indicated by this rate law.

INDEX

Absolute specificity, 176
Acid catalyst, 24, 25
Activated complex, 14, 108
Activated molecule, 108
Activation
   collisional, 103
   enzyme, by metal ions, 191
   volume of, 73
Active sites, 116, 177
Adsorbate, 116
Adsorbent, 116
Adsorption, 115–123
Antoine constants, 138
Antoine equation, 138
Apoenzyme, 176
Arrhenius, 15
Arrhenius equation, 56, 102, 195, 205, 237
Arrhenius plot, 15, 205, 214, 217
Attacking species
   negative, 25
   positive, 24, 25
Autocatalysis, 51–55
Avrami rate law, 208, 210, 214, 215, 232, 234, 236
Avrami-Erofeev rate law. *See* Avrami rate law.
Avrami-Erofeev rate plot, 210

Barrier Penetration, 84
B-E-T isotherm, 121–122
Benson, S. W., 36
Bernasconi, G. F., 65
Berthelot geometric mean, 148
Berzelius, 175
Benzyne, 81, 84
Bodenstein, 110

Boltzmann Distribution Law, 14, 22, 99, 196
Boltzmann population, 133
Born-Oppenheimer approximation, 95
Branching, 114
Brønsted, J. N., 158
Bronsted relationship, 158
Brunauer, 121

Carbocation, 21
Carbonium ion, 21
Catalysis, 24–27, 123–126
   heterogeneous, 115
Catalyst, acid, 24, 25
Chain mechanisms, 19–20, 110–115
Charge dispersion, 143
Charge neutralization, 143
Chemisorption, 116
Christiansen, 111
CNDO, 117
Coats and Redfern equation, 228, 235
Coats and Redfern method, 231, 234, 235, 236, 237
   kinetic analysis by, 224–227
Coats and Redfern plots, 225, 236
Coenzymes, 176
Cofactors, 176
Cohesion energy, 147–150
   density, 137, 147
   solvent, effects on rates, 150–151
Collision complex, 151
Collision theory, 89–93
Collisional activation, 103
Collisional cross section, 90
Combination, direct, 19
Compensation effect, 162–163
Competitive inhibition, 186–187

Complete neglect of differential overlap (CNDO), 117
Comprehensive method, 233–237
Comprehensive nonisothermal kinetic method, 232
Concentration, dependence on, 4–11
Continuous-flow techniques, 77, 78
Contracting area rate law, 202–205
Contracting sphere rate law, 201–202
Contracting volume rate law, 202
Cooperativity
negative, 192, 193
positive, 192, 193
Copper sulfate pentahydrate, dehydration, 210
Coulomb attraction energy, 136
Coulomb's law, 131, 133

Data pairs, method based on four (α, T), 230–231
Data pairs, method based on three (α, T), 228–230
Deaquation-anation of [Co(NH$_3$)$_5$H$_2$O]Cl$_3$, 212–215
Deaquation-anation of [Cr(NH$_3$)$_5$H$_2$O]Br$_3$, 215–216
Debye-Huckel limiting law, 156, 157
Decay region, 198
Dehydration of Trans-[Co(NH$_3$)$_4$Cl$_2$]IO$_3$·2H$_2$O, 216–218
Density, cohesion energy, 137, 147
Differential method, 231–232
Differential scanning calorimetry (DSC), 221–224
Diffusion, 151, 197
Dipole-induced dipole forces, 133
Direct combination, 19
Dispersion forces. See London forces.
Dissociative pathway, 22
DSC methods, 221–224
Eadie analysis, 183–185
Eadie-Hofstee plot, 184
Effective collisions, 93
EHMO, 117
Electron-releasing, 160
Electrostriction, 75

Elementary reactions, 89
Emmett, 121
Energy of vaporization, 137, 147
Energy surface, potential, 93–97
Enthalpy, 221
Enzyme action, 176–177
inhibition of, 185–190
Enzyme activation by metal ions, 191
Enzymes
kinetics of reactions catalyzed by, 178–185
nonregulatory, 192
regulatory, 192–193
Exponential decay, 6
Extended Huckel molecular orbital (EHMO), 117
External pressure, 73, 137
Eyring, H., 98

First-order rate law, 4–6, 124, 180, 200
First-order reactions, 4–6
parallel, 38–39
series, 40–45
Flooding, 70–71
Flow techniques, 77–78
Free-radical mechanisms. See Chain mechanisms.
Freundlich isotherm, 126
Friedel-Crafts reaction, 24

Gases, unimolecular decomposition of, 103–110
Gauss-Jordan condensation, 233
General rate law, 233–237
Group specificity, 176
Grunwald, E., 169

Half-lives, method of, 66–67
Hammett (σ) and (ρ) constants, 155
Hammett equation, 159
Hammett relationship, 159, 161, 162
Hanes-Woolf plot, 184
Hard-soft interaction principle (HSIP), 140–142
Heat of mixing, 146
Henry's constant, 144

Herzfeld, 111
Heterogeneous catalysis, 115
Hildebrand (h), 137
Hildebrand, J., 137
Hildebrand-Scatchard equation, 149
Hill equation, 192
Hill plot, 193
Hinshelwood, C. N., 108
Holoenzyme, 176
Hood, 58

Ideal solutions, 144–150
Ingold, Sir C., 143
Inhibition of enzyme action, 185
    competitive, 186–187
    noncompetitive, 188–189
    uncompetitive, 189–190
Inhibitors, 122–123, 176, 185
Initial rates, 51, 68–70
Initiation step, 20
Insertion reaction, 79
Intermediate, 40
Intermolecular forces, 132–136
Internal pressure, 73, 137, 147
Inverse isotope effect, 83
Ionic strength, effects of, 155–157
Ion(s)
    carbonium, 21
    nitronium, 25
    metal, 176
        enzyme activation by, 191
    solvation of, 139–140
Isokinetic relationship, 162
Isokinetic temperature, 162
Isotherm
    adsorption, Langmuir, 117–121, 124
    B-E-T, 121–122
    Freundlich, 126
Isotope effects, kinetic, 81–85

Kassel, L. S., 108
Keesom, 132
Kelm, H., 76
Kinetic(s)
    analysis by Coats and Redfern
        method, 224–227

data, cautions on treating, 11–13
energy, 93
isotope effects, 81–85
method, nonisothermal,
    comprehensive, 232
Michaelis-Menten, 192
of reactions
    catalyzed by enzymes,
        178–185
    in solids, 195–198
    studies on solids, 210–218
Kirkwood, 135
Kondo, Y., 164

Labelling, 78–81
Laidler, K. J., 166, 167
Langmuir adsorption isotherm, 117–121,
    124
Langmuir, I., 117
Leffler, J. E., 169
Lennard-Jones potential, 136
Lewis acid-base interactions, 140
Lewis acid-base reaction, 21
Lewis acids, 24, 26, 140
Lewis bases, 21, 140
Lewis and Randall rule, 144
LFER, 158–162
Lind, 110
Lindemann, 108
Linear free-energy relationships (LFER),
    158–162
Lineweaver-Burk analysis, 183–185
Lineweaver-Burk plot, 183
Linkage isomerization, 76
Linkage specificity, 176
Liquids, nature of, 131–142
Logarithmic method, 71–73
London attraction energy, 135, 136
London forces, 134, 136, 139, 149
London method, 96

Marcus, R. A., 108
Mares, M., 76
Mass, 221
Maxwell-Boltzmann distribution, 92
Mechanisms, chain, 19–20, 110–115

Metal ions, 176
  enzyme activation by, 191
Method(s)
  based on four (α, T) data pairs,
    230–231
  based on three (α, T) data pairs,
    228–230
  Coats and Redfern, 231, 234, 235, 236,
    237
  comprehensive, 233–237
  differential, 231–232
  of half-lives, 66–67
  kinetic, nonisothermal,
    comprehensive, 232
  kinetic analysis by Coats and Redfern,
    224–227
  logarithmic, 71–73
  London, 96
  Reich and Stivala, 227–228, 231, 232, 234
  TGA and DSC, 221–224
  tracer, 78–81
Michaelis constant, 180, 181, 191
Michaelis-Menten analysis, 178–183
Michaelis-Menten equation, 180
Michaelis-Menten kinetics, 192
Michaelis-Menten model, 192
Molar properties, 145
Molar volume, 73
Molecules, solvation of, 139–140

Negative attacking species, 25
Negative cooperativity, 192, 193
Nicholas, J., 114
Nitronium ion, 25
Noncompetitive inhibition, 187–189
Nonisothermal kinetic method,
    comprehensive, 232
Nonregulatory enzymes, 192
Nucleation, rate laws based on, 207–210
Nuclei, 198
Nucleophiles, 21
Nucleophilic substitution, 21, 154, 167

Palmer, D. A., 76
Parabolic rate law, 199–200

Parallel first-order reactions, 38–39
Parker, 154, 155, 167
Phase space, 98
Physisorption, 116
Poisons, 122–123
Polyani, 111
Positive attacking species, 24, 25
Positive cooperativity, 192, 193
Potential energy surface, 93–97
Pre-steady state period, 77
Pressure
  effects of, 73–76
  external, 73, 137, 147
  internal, 73, 137
Primary kinetic isotope effects, 81
Propagation steps, 20
Prosthetic groups, 176
Prout-Tompkins equation, 193, 205–207
Pseudo first-order reactions, 23
Pseudo zero-order reactions, 11

Ramsperger, H. C., 108
Raoult's law, 144
Rate(s)
  correlations of, with solubility
    parameter, 163–169
  initial, 51, 68–70
  of reactions, 2–4
    factors affecting, in solids, 198–199
  solvation effects on, 151–155
  solvent cohesion energy effects on,
    150–151
  solvent polarity effects on, 142–143
Rate constants, 3
  calculation of, 65–66
Rate-determining step, 3
Rate equation, 2
Rate law(s), 2
  Avrami, 208, 210, 214, 215, 232, 234, 236
  based on nucleation, 207–210
  contracting area, 202–205
  contracting sphere, 201–202
  contracting volume, 202
  first-order, 4–6, 200
  general, 233–237

parabolic, 199–200
second-order, 7–9
zero-order, 9–11
Reaction coordinate, 95
Reaction orders, 35–38
Reaction(s)
deaquation-anation, 212–216
dehydration, 216
elementary, 89
first-order, 4–6
parallel, 38–39
series, 40–45
Friedel-Crafts, 24
insertion, 79
kinetics of
catalyzed by enzymes, 178–185
in solids, 195–198
Lewis acid-base, 21
mechanisms of, 18–24
pseudo first-order, 23
pseudo zero-order, 11
rates of, 2–4
factors affecting, in solids, 198–199
reversible, 45–51
second-order, 7–9
substitution, 21–24, 76
zero-order, 9–11
Regulatory enzymes, 192–193
Reich and Stivala method, 227–228, 231, 232, 234
Reversible reactions, 45–51
Rice, O. K., 108
RRK theory, 108
RRKM theory, 108
Runge-Kutta technique, 233

Sapunov, V. N., 61
SCF, 117
Schmid, R., 61
Second-order case, first-order in two components, 31–35
Second-order rate law, 7–9
Second-order reactions, 7–9
Secondary isotope effects, 85
Self-consistent field (SCF), 117

Series first-order reactions, 40–45
Sigmodial plot, 55, 208
Significant structure theory, 131
Sintering, 26, 198
Slater, 135
Slow step, 3
Solids
diffusion in, 196–197
factors affecting reaction rates in, 198–199
kinetics of reactions in, 195–198
sintering, 26, 198
Solubility parameter, 136–139
correlations of rate with, 163–169
Solutions, ideal, 144–150
Solvation
effects on rates, 151–155
of ions and molecules, 139–140
Solvent cohesion energy effects on rates, 150–151
Solvent polarity, 164
effects on rates, 142–143
Species, attacking
negative, 25
positive, 24, 25
Specificity
absolute, 176
group, 176
linkage, 176
stereochemical, 176
Steady-state approximation, 45, 107, 109, 111
Stereochemical specificity, 176
Stopped-flow technique, 77, 78
Substitution reactions, 21–24, 76
Substrate, 175
Sumner, J. B., 175
Surface, energy, potential, 93–97

Taft equation, 161, 162
Taft, R. W., 161
Teller, 121
Temperature
effects of, 14–18, 55–61
isokinetic, 162

Termination steps, 20
TGA methods, 221–224
Thermogravimetric analysis (TGA), 221–224
Third-order rate law, 36–38, 218
Topochemistry, 198
Tracer methods, 78–81
Transient period, 77
Transition state, 14
    theory, 98–103
Transparency, 84, 96

Uncompetitive inhibition, 189–190
Unimolecular decomposition of gases,
    103–110

van der Waals equation, 147
van der Waals forces, 147, 148
Van Laar equation, 148
Vaporization, energy of, 137, 147
Variational transition state theory, 102
Volume of activation, 73

White, M. G., 117, 120

Young, D. A., 205

Zero-order rate law, 9–11
Zero-order reactions, 9–11, 124, 180